SpringerBriefs in Archaeology

SpringerBriefs in Underwater Archaeology

Series Editor

Annalies Corbin, Columbus, USA

Books in the SpringerBriefs in Underwater Archaeology series, published in cooperation with the Advisory Council on Underwater Archaeology (ACUA), will address critical contemporary problems and illustrate exemplary work in maritime cultural heritage in countries around the globe. The series will take a broad view of the theory, concepts, issues and complexities associated with the management and protection of maritime cultural heritage sites in the 21st century. The book series will be concerned with:

- submerged cultural resources ranging in age from the most ancient human history through the modern era
- maritime cultural heritage, understood as not just archaeological sites but also the stakeholders who inevitably constitute diverse and multi-scalar sectors of society
- management that transcends adherence simply to regulation but that looks forward to the needs of the future
- emerging technologies that will transform the ways in which archaeological research is conducted and archaeological resources, both sites and landscapes, are understood and interpreted
- site stewardship and dissemination of information that serves the needs of the archaeological field and the public at large.

All volumes will be stringently peer-reviewed, first by the ACUA series editor and associate editors, and upon passing that level, two outside reviewers familiar and qualified to make constructive comments regarding the manuscript.

Charlotte Jarvis
Editor

Threats to Our Ocean Heritage: Deep Sea Mining

2021 2030 United Nations Decade of Ocean Science for Sustainable Development

The Ocean Foundation

Springer

Editor
Charlotte Jarvis
The Ocean Foundation
Washington, DC, USA

ISSN 1861-6623 ISSN 2192-4910 (electronic)
SpringerBriefs in Archaeology
ISSN 2625-2562 ISSN 2625-2570 (electronic)
SpringerBriefs in Underwater Archaeology
ISBN 978-3-031-98237-8 ISBN 978-3-031-98238-5 (eBook)
https://doi.org/10.1007/978-3-031-98238-5

This work was supported by Lloyd's Register Foundation (GBY\100046).

This Springer imprint is published by the registered company Springer Nature Switzerland AG
The registered company address is: Gewerbestrasse 11, 6330 Cham, Switzerland

Foreword

Threats to Our Ocean Heritage: Deep Sea Mining is the third component of a trilogy of books initiated by The Ocean Foundation published by Springer that focus on the risks to the ocean's natural and cultural heritage, noting that the zones at risk should extend to include seas, lakes and other aquatic places.

Combined, the volumes *Potentially Polluting Wrecks*, *Bottom Trawling*, and *Deep Sea Mining* are raising international awareness of the interaction of the physical, biological, and chemical risks to heritage in the ocean. Inadequate operating standards and legal safeguards are also a factor and increase the overall risk. All aspects of the associated risks are well covered and discussed in the three volumes and especially here for deep sea mining (DSM).

The definition of deep sea is below a depth of 200 metres, which gives the impression of being somehow protective of what lies within these waters. Although reflective of the past, it does not take into account the ever-advancing development of remote technologies that enable much of the ocean's floor, reaching thousands of metres, to be accessible. Two hundred metres is even within the range of human intervention; consequently whether manned or unmanned interventions, there is an enhanced level of risk and impact on natural and cultural Ocean Heritage.

Aside from the consequences of the impacts mentioned above, perhaps the more prescient risk is the inadequacies of not fully implementing existing international regulations. 'Rise Up: a blue call to action'[1] (UNOC 3) calls for better management of the ocean including a 'ban on Deep-Sea Mining and a moratorium', though others want a total ban and some actors want unregulated access. Improving regulations, along with management best practices, would provide greater protection to natural and cultural heritage, particularly required in places beyond the limit of state jurisdictions.

The case made in support of DSM is compelling, claiming that the recovery of minerals in the deep ocean is considered essential for decarbonising international

[1] https://riseupfortheocean.org/campaigns/the-un-ocean-conference-2025/

industries and help mitigate the extremes of climate change. The counter is at what cost in terms of damage to natural ecosystems and underwater cultural heritage?

The focus has been, understandably, largely on the impact on natural heritage. However, underwater cultural heritage is also at significant risk and the two types of heritage are in many instances linked. Shipwrecks are a sanctuary for many species; the physical remains of shipwrecks are often the bedrock for marine ecosystems. A number of shipwrecks are mentioned in this volume such as the polar research vessel *Endurance*. Its apparent state of preservation is outstanding; the approach to the management of the site, so far, is exemplary and inclusive of natural and cultural heritage considered of mutual relevance and importance. The first step has been to pause all interventions on the site to allow time for the development of a Conservation Management Plan. An important component is the extension of the exclusion zone to encompass all cultural material and in parallel develop research projects that will determine the ship's physical, biological, and chemical attributes, and identify associated threats to the site, both natural and anthropogenic. This approach should act as a benchmark for future protection and management of the deep sea. As was made clear in ICOMOS-ICUCH's intervention during Spain's side-event meeting of the International Seabed Authority (March 2024), there is an imperative need for a pause to enable research to better understand the impact of DSM by encouraging a precautionary approach to the management of the natural and cultural heritage within the ocean.

Buenos Aires, Argentina
April 2025

Christopher J. Underwood

Preface

The deep sea is a place of wonder and fantasy, brimming with other-worldly life—ancient, awe-inspiring, and utterly unlike anything we are familiar with as land dwellers. Thanks to the growing number of expeditions and the generosity of the scientists and investors who openly share the results of those journeys to one of the last largely unexplored parts of our planet, our understanding of the treasures of the deep is growing quickly.

The volume and interconnectedness of the deep sea's plant and animal life—great and small, visible and microscopic, adorable and downright alien—are key to two of the ocean's critical services to our planet. As part of the greatest migration on Earth, every day, billions upon billions of animals migrate between the depths and the surface in a circadian dance that affirms the deep ocean's role as a vital source of both food security and climate regulation.

There is so much more to learn about this migration, along with a multitude of other ocean forms and processes. What we now know about the chemistry of hydrothermal vents has upended earlier theories that photosynthesis was the only source of life energy on our planet, and there is still much more research to be done. Just last year, scientists confirmed that polymetallic nodules on the floor of the deep abyss actually produce oxygen—a conclusion so unexpected that it took several years for them to believe their own data.

Although in one sense we remain in the early stages of our learning journey, for millennia, coastal cultures have understood, respected, and revered the ocean's abundance, grace, and connectivity. In recent decades, as a more industrial and capitalist exploitation of the ocean has prevailed, it has become starkly clear that in our quest to generate private profit, we are capable of causing damage that is both extensive and in some cases irreversible.

Deep sea mining is the latest in a string of such assaults. Unless our governments press 'pause' with a decision for a moratorium, this emerging industry could take off, risking a cumulative cost to ocean and planetary health, and to future generations, which far outweighs the likely benefits to all but a few. This book does an important service of weaving together the legal, scientific, and conservation threads of the deep sea mining story with the historical and social perspectives of underwater cultural heritage. May it succeed in opening new lines of dialogue and action across these communities and beyond, and in so doing, help to secure the long-term health of the deep ocean for us all.

Deep Sea Conservation Coalition
Amsterdam, Netherlands

Sian Owen

Acknowledgements

Deep sea mining was not something that I had heard of when I started working in the field of underwater cultural heritage a few years ago. But I was introduced to the issue when I joined The Ocean Foundation. I quickly became engrossed in the issue and concentrated on raising awareness of the threats DSM poses to cultural heritage. Deep sea mining became a focus for me and I was brought into a project which aimed to show destruction to heritage from different activities, such as bottom trawling. The multitude threats of DSM presented a unique opportunity to campaign for heritage protections before a damaging industry began. I was hooked.

Thanks to Bobbi-Jo Dobush, especially for her expert guidance and assistance with not only this book, but also bringing UCH to the ISA and DSM discussions generally. I have learned so much from you and I am so grateful. Thanks to the many individuals at other organsiations, such as the Deep Sea Conservation Coalition, for their assistance with this volume and research. It would not have been possible without advice from experts in the field and also from the countless maritime archaeologists who have advised on the book and helped me work through sections. We may be living in scary times for the ocean's health and safety, but it is inspiring to work alongside so many individuals who are fighting to protect Ocean Heritage.

This is the final book in the initial phase of the 'Threats to Our Ocean Heritage Project', for which I am privileged to be the project leader. The first two books, *Threats to Our Ocean Heritage: Bottom Trawling* and *Threats to Our Ocean Heritage: Potentially Polluting Wrecks*, have been a great success, and I want to thank everyone involved in the project who helped make these three books happen, especially Michael Brennan, Annalies Corbin, and Christi Lue. None of these volumes would have been possible without the research and expertise of so many individuals and organisations in the fields of maritime archaeology, cultural heritage, natural heritage, heritage law, and more.

This book, and the entire project, would not be possible without the generous support from the Lloyd's Register Foundation Heritage and Education Centre. Thanks to the director, Alex Stitt, for supporting the project and helping design it in the crucial early stages and to the rest of the team at LRF for their help and support.

Without the help of Louise Sanger and Lydia Wooley this project would not be what it is today.

Thanks to the team at The Ocean Foundation (TOF) who have supported and made this project possible over the last 2 years. The Ocean Foundation is the only community foundation for the ocean dedicated to reversing the trend of destruction of ocean environments around the world, and it is one of the few ocean NGOs that focuses on both natural and cultural heritage, making it uniquely situated to bring this project to light. Thanks to the president, Mark Spalding, for championing underwater cultural heritage and investing in research that makes projects like this possible. Thanks to the rest of the cultural heritage team at TOF, Ole Varmer and Benjamin Ferrari, for their support and assistance bringing this project together.

I'm thrilled that this project has been endorsed under the UN Decade of Ocean Science for Sustainable Development and is part of the Cultural Heritage Framework Programme. Thanks to the team we work with there. We are also fortunate to have the support of the International Committee on Underwater Cultural Heritage (ICUCH) within the International Committee on Monuments and Sites (ICOMOS) and the Advisory Council on Underwater Archaeology (ACUA). My particular thanks go to the reviewers of this book, in all its many stages, whose insights and suggestions have helped make it a stronger contribution to the field and to the editing team at Springer. And, finally, thanks to my friends, family, and colleagues for their support along the way.

Charlotte Jarvis

Contents

List of Abbreviations

ABMT	Area-Based Management Tools
ABNJ	Areas Beyond National Jurisdiction
AIMA	Australasian Institute for Maritime Archaeology
AIS	Automatic Identification System
AUV	Autonomous Underwater Vehicle
BBNJ	Agreement Under the United Nations Convention on the Law of the Sea on the Conservation and Sustainable Use of Marine Biological Diversity of Areas Beyond National Jurisdiction
BOEM	Bureau of Ocean Energy Management (US)
CCZ	Clarion Clipperton Zone
CHIA	Cultural Heritage Impact Assessment
CoP	Conference of the Parties
CRC	Cobalt-Rich Crusts
CS	Continental Shelf
DSM	Deep Sea Mining
DSMV	Deep Sea Mining Vehicle
EBSA	Ecologically or Biologically Significant Area
EEZ	Exclusive Economic Zone
EIA	Environmental Impact Assessment
ESIA	Environmental and Social Impact Assessment
EU	European Union
HE	Historic England
ICH	Intangible Cultural Heritage
IPLC	Indigenous Peoples and Local Communities
ISA	International Seabed Authority
ITLOS	International Tribunal for the Law of the Sea
IUCN	International Union for the Conservation of Nature
LOSC	Law of the Sea Convention (United Nations)
LRF	Lloyd's Register Foundation
MAANZ	Maritime Archaeological Association of New Zealand
MAG	Magnetometer

MBES	Multibeam Echosounder
MGR	Marine Genetic Resource
MMO	Marine Management Organisation
MMS	Minerals Management Service (US, now BOEM)
MPA	Marine Protected Area
NGO	Non-Governmental Organisation
NM	Nautical Miles
NOAA	National Oceanic and Atmospheric Administration (US)
PMN	Polymetallic Manganese Nodules
PPW	Potentially Polluting Wreck
REMP	Regional Environmental Management Plan
ROV	Remotely Operated Vehicle
RULET	Remediation of Underwater Legacy Underwater Threats
SBP	Sub-Bottom Profiler
SDG	Sustainable Development Goal
SEA	Strategic Environmental Assessment
SMS	Seafloor Massive Sulphides
SSS	Side-Scan Sonar
TEK	Traditional Ecological Knowledge
UCH	Underwater Cultural Heritage
UN	United Nations
UNESCO	United Nations Educational, Scientific and Cultural Organization
UNGA	United Nations General Assembly
WWII	World War II

List of Figures

Chapter 1
Introduction

Charlotte Jarvis

Keywords Underwater cultural heritage · Deep sea mining · Deep seabed mining · Deep-sea mining · Intangible cultural heritage · International seabed authority · Underwater archaeology

Underwater cultural heritage (UCH) is integral to both cultural heritage and natural heritage, thus making it a shared **Ocean Heritage**. The stories of our societies and our ancestors are wrapped in intangible connections with the ocean and preserved on the seafloor as artefacts, shipwrecks, and remains of those lost or buried at sea. Deep sea mining (DSM) threatens this Ocean Heritage and has the potential to destroy crucial pieces of our past and disrupt our deep cultural connections to the ocean. This book is designed to highlight the potential threat of DSM, covered here mostly in international waters and as regulated by the International Seabed Authority (ISA), and make ocean professionals across many different fields aware of those dangers and of the work being done to help mitigate the possible launch of DSM operations. It does not aim to be the final word on the issue, but rather start a discussion and encourage scientists and cultural leaders to engage and work together.

When beginning a project that sheds light on threats to our Ocean Heritage, we may simply look at those threats that currently exist, as has been done already in *Threats to Our Ocean Heritage: Bottom Trawling* (Jarvis, 2024) and *Potentially Polluting Wrecks* (Brennan, 2024). However, there are also potential dangers to Ocean Heritage—DSM is one of these, but one that many UCH professionals might not yet have considered.

Deep sea mining is a 'potential commercial industry with a goal to mine mineral deposits from the ocean floor, in the hopes of extracting metals such as manganese,

C. Jarvis (✉)
The Ocean Foundation, Washington, DC, USA

C. Jarvis (ed.), *Threats to Our Ocean Heritage: Deep Sea Mining*,
SpringerBriefs in Archaeology, https://doi.org/10.1007/978-3-031-98238-5_1

copper, cobalt, and nickel' (Dobush & Warner, 2024, p. 4). The deep sea, defined as the ocean below 200 metres, was previously considered uninhabited and largely unimportant (if considered at all), but that has changed over the last half century as deep-water exploration has increased. It is an enormous spread of ocean (covering 65% of the Earth's surface and more than 95% of the biosphere) with a biodiversity akin to tropical rainforests (Danovaro et al., 2010). The deep ocean provides many critical ecosystem services, including fish and shellfish for food; products from organisms that can be used for medicines; climate regulation; and historical, cultural, social, educational, and scientific value for people worldwide (Thurber et al., 2014; Le et al., 2017; Armstrong et al., 2012).

The deep sea is often referred to as earth's final frontier, but its inaccessibility means it receives relatively less attention than other ocean areas. The nineteenth century brought some of the first serious scientific interest in the ecosystem and what lived in it (Cuyvers et al., 2018). As early as the 1870s, researchers on HMS *Challenger* collected samples of polymetallic nodules (fist-sized mineral accretions) (Cuyvers et al., 2018). By the 1960s, these nodules were studied in more detail and the ocean was cited as a potential site to mine for the minerals inside. By the 1970s, companies were beginning to test mining methods, but the idea was shelved when metal demand dropped (Cuyvers et al., 2018).

But the deep sea has been important and vital to humans long before scientists entered the picture. Many cultural groups around the globe have creation stories linked to the ocean, and especially the deep ocean. The Haida, an indigenous group based for at least 12,500 years in the Haida Gwaii archipelago off the north Pacific coast of Canada, link the supernatural being SGaan Kinghlas, to the formation of a seamount within Canada's Exclusive Economic Zone (EEZ) (Fedje & Mathewes, 2005). In Africa, some cultures see the sea as their ancestral spirits' home and the ocean as a sacred place, with respect for it increasing with its depth. In South Africa, the Nguni-speaking diviner-healers cite travel underwater with spirits as a source of their power. They refer to the sea as '*Komkhulu*' (The Great Place) and some Zulu see it as a place where the water divinities live (Bernard, 2010). The deep sea has also had long-held significance in cultures all over as a habitat for marine species, such as whales, sharks, or turtles, which hold deep cultural importance (see Chap. 8, this vol. for more details).

DSM has the potential to cause major disturbances to the marine environment, including direct damage to the benthic fauna, habitat destruction, pollution from sediment plumes and wastewater discharge, and non-stop noise and light pollution across the water column (Weaver et al., 2018; Jones et al., 2018). Scientists expect that DSM would have devastatingly wide and irreversible impacts on microorganisms, fisheries, food chains, cultural traditions, and more by impacting the seafloor and water column (Jaeckel et al., 2023; Dobush & Warner, 2024) (Fig. 1.1).

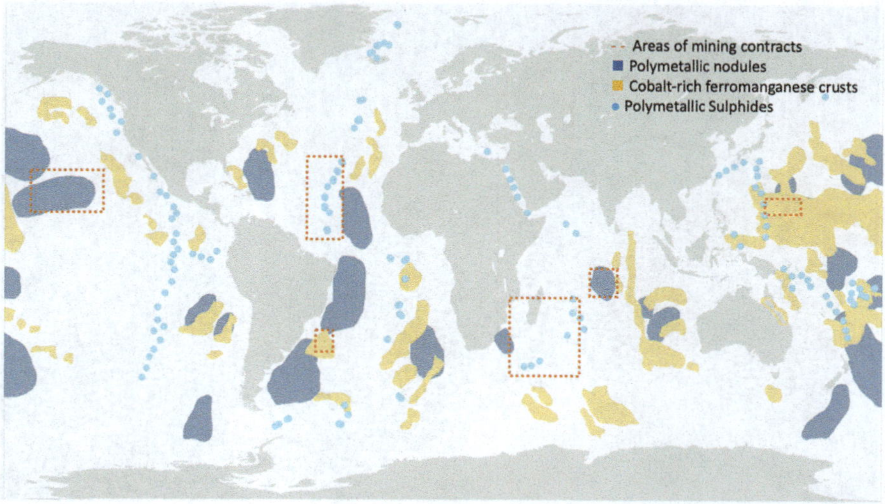

Fig. 1.1 Map of the elements for proposed mining as well as the zones in which the ISA has issued contracts. The Clarion Clipperton Zone, left side of the map, has been the most sought-after with 17 contracts issued for that area alone. (Source: Created by author, 2024. Based on information from the International Seabed Authority)

1.1 What Are Seabed Minerals?

Minerals are deposited and formed on and within the seabed in various ways, including as polymetallic nodules, polymetallic massive sulphides, and cobalt-rich ferromanganese crusts. The main areas of mineral deposits that commercial mining is focused on are located on the abyssal plains (for nodules), hydrothermal vents (for sulphides), and seamounts (for crusts).

Polymetallic nodules, sometimes called manganese nodules, occur primarily at depths below 4000 m and are found on the seafloor rather than embedded in hard crust (Fig. 1.2). They range in size from 4 to 10 cm in diameter and take millions of years to grow to the size targeted by mining (Cuyvers et al., 2018; Miller et al., 2018). Companies seeking to start the industry have a focus on minerals located in the Clarion Clipperton Zone (CCZ) in the Pacific, in the deep abyssal plain between Hawaii and Central America, as this area has a high concentration of nodules. DSM proponents point to the high concentration of multiple sought-after minerals in each nodule (Muñoz-Royo et al., 2021). The nodules 'provide the rare commodity of hard substrate' on the seafloor and are used as important habitat for immobile animals such as sponges, corals, or anemones on the abyssal plain to grow on (Stratmann et al., 2021, p. 1).

Polymetallic massive sulphides, or seafloor massive sulphides, are formed at hydrothermal vents where seawater meets magma (Fig. 1.3). The vents are biomass hot-spots as species use the chemical-rich liquid escaping from the vents as a source of energy rather than sunlight (chemosynthesis) (Van Dover et al., 2018). These vents 'provide habitat for novel faunal assemblages that have altered our views of

Fig. 1.2 A field of manganese nodules. (Source: NOAA Office of Ocean Exploration and Research)

Fig. 1.3 Active smoker chimneys precipitating iron, copper, and zinc sulphides. (Source: NOAA Office of Ocean Exploration, Pacific Ring of Fire 2004 Expedition)

the primary energy sources and origins of life and exhibit substantial local and regional variation in structure and connectivity' (Smith et al., 2020, p. 853). Active hydrothermal vents 'host high-temperature (350 °C) "black smokers" that discharge metal-rich fluids' (van Dover et al., 2018, p. 21). The metals that precipitate form the polymetallic (especially copper and zinc) sulphides which are the inactive vent sites targeted by the DSM industry. The 'general perception of inactive sulphide deposits as biologically depauperate' means there has not been much exploration or biological characterisation of them, though there are some studies that show corals, sponges, or other species at inactive sites (Levin et al., 2016, p. 251; Van Dover et al., 1990; Erickson et al., 2009; Boschen et al., 2016).

Cobalt-rich crusts, or ferromanganese crusts, form on the slopes of seamounts and other underwater formations (Miller et al., 2018) (Fig. 1.4). The definition of seamounts can vary, but in the deep sea, they generally are 'isolated topographic elevations with summit depths at least 100 m above the seafloor' and are often volcanic in origin (Rogers, 2018, p. 137; Yesson et al., 2011). They support diverse ecological communities which 'take centuries or more to grow and are home to long-lived animals with complex interdependent relationships' (DOSI, 2023b, p. 2).

The metals targeted by DSM proposals are largely posited by mining companies as providing the building blocks necessary to decarbonise or move away from fossil fuels through batteries for things like electric vehicles (EVs) or as 'critical minerals' for military or other potential uses. The depletion of terrestrial deposits has also been cited as a reason to push forward with DSM (Filho et al., 2021). However, some new and next generation batteries use metals that are not targeted by DSM. As electric vehicle battery technology advances, there is a shift towards replacing batteries dependent on cobalt, nickel, and manganese with ones that require fewer minerals that need to be mined from the sea, such as lithium iron phosphate or sodium-ion batteries (Everett et al., 2023). The growth of lithium iron phosphate (LFP) batteries, which do not require minerals targeted by DSM, has been rapid, capturing a

Fig. 1.4 A ~4200-m high seamount mapped during a NOAA expedition. (Source: NOAA Office of Ocean Exploration and Research, Mountains in the Deep: Exploring the Central Pacific Basin)

significant portion of the global EV battery market. Additionally, investment in the circular economy is directing focus to recycling and reuse of existing materials.

At the time of writing, commercial DSM in international waters has not begun, and there are still many unknowns as to how commercial scale mining could proceed (see Appendix I for ISA-issued exploration contracts and zones). Mining polymetallic nodules would require a machine, some companies' technology resembles a three-story-tall tractor, which is taken to the seafloor to take up the top layer of sediment containing the nodules with the targeted minerals; this process would also recover any other items on the seafloor such as rocks, flora, and small fauna (Fig. 1.5). These materials are transported up to a surface vessel where the slurry is processed and unneeded material is returned to the water via a discharge plume. The impacts of this method have been likened to strip mining. Another company has proposed a technology which would use artificial intelligence (AI) to pluck each nodule individually from the seafloor and posit this method as being less damaging to the seafloor habitats (Impossible Metals, 2024), despite the nodules themselves being critical habitats.

1.2 The Importance of Our Ocean Heritage

1.2.1 Underwater Cultural Heritage

Recognition of the potential devastation of UCH by DSM practices lags far behind the research conducted on natural heritage. The impacts of DSM could encompass both destruction of physical objects on the seafloor as well as damages to cultural practices and connections to the ocean. In the Pacific especially, DSM is related to the island environment because 'the ocean is at the heart of one's identity' and 'integral to the core identity of Islanders' (Tilot et al., 2021, p. 4). Other important elements of UCH in the Pacific include canoe and wayfinding practices, which are integral to cultural identity in Oceania. Tangible evidence of this is rare, but includes spiritual associations with natural heritage, such as shark calling (Jeffery et al., 2021). Additionally, vessels and shipwrecks represent important artefacts of human history and are the final resting place for many that have sunk in the open water areas of the high seas, including those from World Wars (Chap. 6, this vol.) and the Atlantic Slave Trade (Turner et al., 2020).

1.2.2 Natural Heritage[1]

Deep sea mining presents many risks to the cultural environment, but it is also anticipated to cause impacts at all levels of ocean ecosystems. The nature of impacts foreseen from DSM are numerous and can include direct or more indirect impacts.

[1] Thank you to Maila Guilhon for her input in this section.

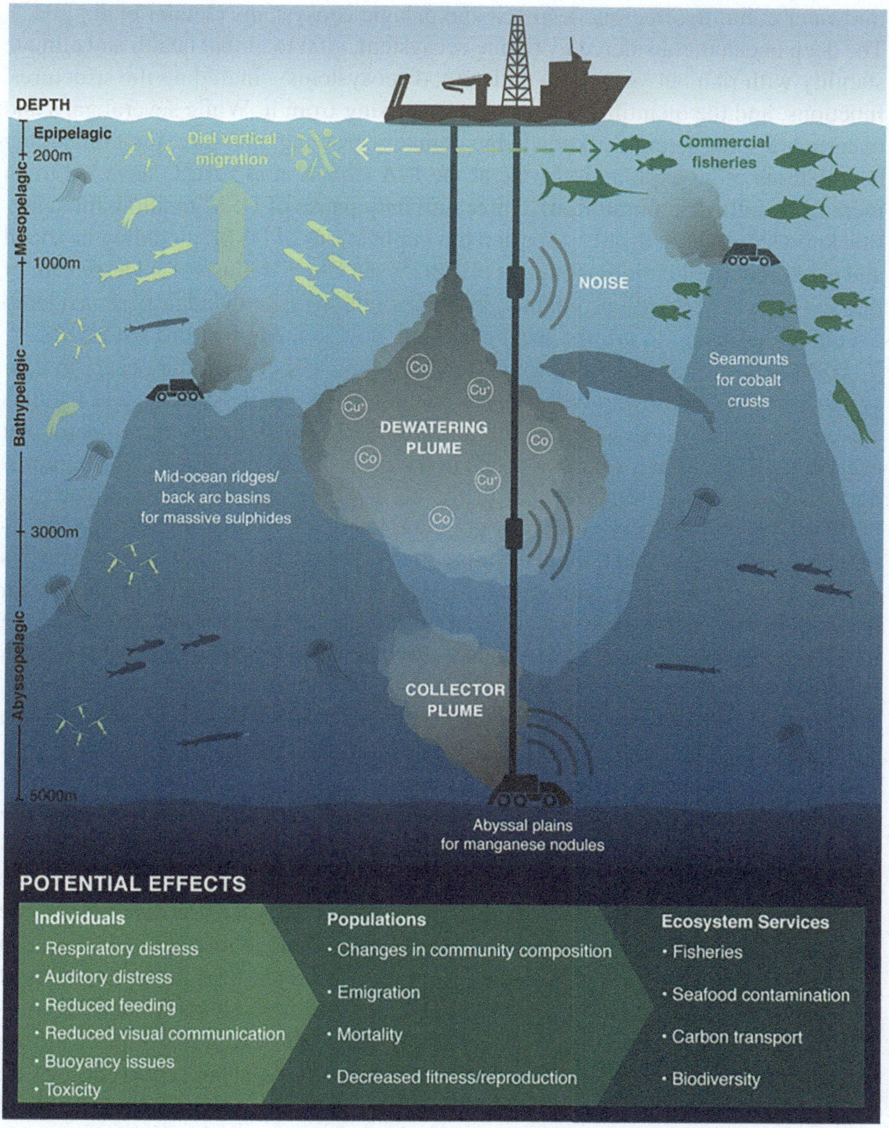

Fig. 1.5 Potential impacts of mining-generated sediment plumes (organisms and plume not to scale). (Source: © Drazen et al., 2020, PNAS. Graphic Artist: Amanda Merritt. Reprinted with permission from artist)

Direct impacts would be those that result from the extraction of the mineral structures and the destruction of biodiversity that live on or around it, the destruction or modification of habitats (Jenner et al., 2023). Indirect impacts are more diffuse, uncertain, and difficult to predict. These include, for instance, noise and light pollution and the formation of sediment plumes arising from machinery activity. Other indirect impacts can result from the discharge of return-water through pipes in the midwater column, affecting deep-sea and pelagic ecosystems (Jenner et al., 2023). The deep ocean is considered a pristine ecosystem, vital to global health and climate stability with rich but yet poorly understood ecosystems—including the structures, functions, and the quantification of benefits arising from it. While several gaps and uncertainties on scientific knowledge remain, growing pressure for concluding negotiations of an exploitation code at the ISA (Singh et al., 2021) have led to an increasing call for a moratorium, a precautionary pause of DSM to allow for scientific knowledge about the deep-sea and the implications of DSM, including its social and economic implications (Singh et al., 2025; see Chap. 4, this vol.). Despite all of the above, a key question remains, at what point will knowledge be considered enough or sufficient to inform decisions if and how to move forward with DSM?

DSM activities are foreseen to take place kilometres deep in the ocean, in ecosystems largely unknown to science. In general, deep-sea ecosystems and habitats are characterised by slow dynamics and processes. In practice, that means that everything happens at a very slow pace in the deep ocean, including growth, reproduction and recovery. Due to that, in the occasion of impact, for example, from DSM activities, full ecosystems may not completely recover. As an example, a perturbation experiment that took place more than 30 years ago in the Pacific Basin has shown very low recovery rates (Simon-Lledó et al., 2019). Potential impacts from DSM are not restricted to the ocean floor. In fact, the impact of organism in the deep midwater (from 200 m to 5 km deep) has raised concern to scientists, given that this environment represents 'more than 90% of the biosphere, contain[s] fish biomass 100 times greater than the global annual fish catch, connect[s] shallow and deep-sea ecosystems, and play[s] key roles in carbon cycling, nutrient regeneration, and provisioning of harvestable fish stocks' (Drazen et al., 2020, p. 17456).

Deep-sea ecosystems provide a number of services and societal benefits (Le et al., 2017; Thurber et al., 2014; Tilot et al., 2021). Deep-sea ecosystems, like seamounts, play a very important role in the maintenance of important fishing stocks (see Gianni, 2004; Ramierez-Llodra, 2020). Consequently, an increase in pressure over seamount ecosystems through mining, for instance, could lead to impact in other sectors of commercial activities. Currents distribute heat around the planet which drives weather systems. Nutrients, vital to marine food chains, are regenerated at the ocean's surface by a process of upwelling, that bring new nutrients from deeper waters up to the surface. Moreover, the deep-sea is the largest carbon sink on earth with microorganisms and sediments helping to sequester some 25% of all carbon dioxide emissions since the mid-1990s (Hilmi et al., 2023; Gruber et al., 2019; Watson et al., 2020), having an important role in regulating the planet's temperature. The deep ocean also houses many migrating species, including whales, which have cultural value for many indigenous communities (Tilot et al., 2021). Finally, minerals structures such as polymetallic nodules serve as habitats by providing hard-subtract for organisms to fix on.

As the minerals found on the seafloor derive from different process of formation and deposit, foreseen technics to extract such minerals are very different. Although such technologies are yet under development, manganese nodules are likely to be dredged off the seafloor, whereas SMS and cobalt crust extraction will utilise rock-cutting technology. Based on such, the main expected impacts into deep-sea ecosystems and biodiversity are listed by the IUCN (2022) as follows:

Disturbance of the seafloor The most direct impact of DSM and digging and gauging of the ocean will result in fragmentation or loss of ecosystem structure and function.

Sediment plumes Fine sediments will be disturbed on the seafloor creating plumes of suspended particles exacerbated by mining ships discharging wastewater at the surface. Such plumes, potentially dispersed for hundreds of kilometres could smother animals, harm filter-feeding species, and block visual communication.

Pollution Species such as whales, tuna and sharks maybe affected by noise, vibrations and light pollution caused by mining equipment and surface vessels, as well as potential leaks and spills of fuel and toxic products.

Recovery dynamics Several studies highlight that the deep seabed may not naturally recover from the impacts of DSM activities (Amadi & Mosnier, 2023; De Jonge et al., 2020).

Historically, deep sea areas have been understudied (Costello et al., 2010) with only 0.00001% investigated (Danovaro et al., 2020). Although over the years there has been an increase in deep-sea research and knowledge, the deep ocean remains an understudied environment that we know very little about. From a single research expedition at a contract region at the CCZ, which holds 17 exploration contracts at the ISA, of the more than 5000 species sampled—88% of the species examined were previously undescribed (Rabone et al., 2023). As another example, a recent discovery has demonstrated that polymetallic nodules may have the capacity to produce oxygen through a process of electrolysis (Sweetman et al., 2024). These are good examples reflecting how limited our knowledge about the deep sea and its biodiversity and structure is, consequently, how we know very little about how it works. Such extent of uncertainties requires the adoption of a precautionary approach into decision-making at the ISA.

Two recent studies provide an assessment of how available information on deep sea ecology matches up to the challenge of making wise judgements in the face of complexity and inherent uncertainty. The first addressed, in broad terms, the requirements of widely accepted standards for environmental impact assessment and associated management plans. The conclusion was that for the majority of matters that would be considered, there is neither adequate baseline data nor reliable information on impacts (Amon et al., 2022). A White Paper produced by the World Economic Forum (Yeh & WEF, 2022) referenced this work and extended the question to whether, given the current state of knowledge, DSM minerals can be included in responsibly sourced supply chains. The conclusion again pointed to very large knowledge gaps and also weakness in stakeholder engagement. It was found that the potential effects that are currently the least predictable are the ones that most directly affect people and planet. Both studies touched on what effort might be required to close these gaps. Expert estimates ranged greatly but few suggested less than ten years of highly concentrated work, and many predicted much longer timescales.

1.2.3 Underwater Cultural Heritage as Both Natural and Cultural

The Pacific Ocean, in particular, serves as an important case study in which the natural and cultural heritage of the ocean is linked and of 'profound cultural, social, spiritual, and economic significance to Pacific Island communities' (Tilot et al., 2021, p. 2). This region has a high number of potentially polluting wrecks (PPWs), shipwrecks 'containing a cargo or a large volume of its own fuel that remains within the wreck and has the potential to pollute due to petroleum cargoes or bunkers' (Brennan, 2024, p. 1; further information see Chap. 6, this vol.). WWII especially left a large footprint in the Asia-Pacific region comprising thousands of shipwrecks, plane wrecks, and other items from both sides (Jeffery et al., 2021). Many of these World War wrecks sit in areas that are already a focus of seabed mining, as seen in Chap. 6. The risk of hitting one of these already fragile vessels and causing a massive oil leak is high and must be considered.

1.3 Societal Impacts of Deep Sea Mining

There are also potential risks for threats to human lives and worker safety associated with deep sea mining as well as socio-economic problems and impacts to coastal livelihoods. Large-scale development of DSM may cause a loss of access to coastal sites of cultural significance or those that are vital for fishing, tourism, or other economic activities (UNEP FI, 2022). Additionally, there is a risk of impact to human health through disruption to marine biodiversity which may impact food safety, security, and an acceleration of global climate change (Hamley, 2022).

Many studies have pointed to alternative sources for the minerals targeted in seabed mining. Miller et al. (2018) summarise the options as 'substituting metals in short supply, such as rare earths, for more abundant minerals with similar properties; landfill mining; and collection and recycling of components from products at the end of their life cycle' (p. 19). They also suggest promoting responsible e-waste recycling and a shift to a focus on reducing consumption. The European Academies Science Advisory Council (EASAC) found that 'the pressure for mining is driven by industry and economic interests rather than demands from the transition to a green economy' (EASAC, 2023, p. 17).

There are countless uncertainties regarding DSM. Although many companies have proposed technology to mine the minerals, at the time of writing, none have publicly published outlined measures for the processing, storage, transportation, port transfers, or land infrastructure to conduct the operation. But any transport will likely have the same negative impacts as seen in other forms of maritime transport, including 'marine pollution, atmospheric emission, underwater noise and physical disturbance' (UNEP FI, 2022, p. 26). Additionally, equality of revenue and benefit sharing has not been guaranteed. Some ISA discussions have focused on financial sharing, but no consensus has been reached (Pickens et al., 2024). 'Rights of Nature', a theory that cites the inherent rights as associated with the ecosystem and species similarly to the concept of human rights, have not been well considered either (Miller et al., 2021). Miller and colleagues argue that while 'the concept of giving rights to nature might seem new to many in the industrialised West, the concept is not unusual to some Indigenous communities' (Miller et al. 2021, p. 5). Ecuador, for example, includes Rights of Nature in its constitution and this could be applied to the ocean alongside a precautionary principle (see Chap. 2, this vol.).

At the time of writing, there are voices speaking out against DSM in all sectors. Businesses and financial institutions including Apple, Microsoft, Google, Samsung, SAP, Phillips, ABN AMRO, BBVA, Cooperative Banking, Lloyds Banking Group, NatWest, Credit Suisse, Standard Charter Bank, Triodos Bank, and Patagonia have voiced support for a DSM moratorium or pledged to cut deep sea metals out of their supply chains.[2] Some in the automotive industry have also come out against DSM including BMW, Volvo, Volkswagen, Daimler, General Motors, Renault, and Ford.[3]

[2] For an overall discussion on companies against DSM see: WWF. (2021, December). *Business Backlash Against Deep-Sea Mining Grows*. https://wwf.panda.org/wwf_news/?4541466/Business-Backlash-Against-Deep-Sea-Mining-Grows; Greenpeace. (n.d.). *Race to the Top*. Stop Deep Sea Mining. Retrieved January 22, 2024, from https://www.racetothetop.app; Deep Sea Conservation Coalition. (n.d.). Voices calling for a moratorium: Companies and finance. Retrieved July 8, 2024, from https://deep-sea-conservation.org/solutions/no-deep-sea-mining/momentum-for-a-moratorium/companies-and-finance/; and Reid, H. (2021, March 31). Google, BMW, AB Volvo, Samsung back environmental call for pause on deep-sea mining. *Reuters*. https://www.reuters.com/business/sustainable-business/google-bmw-volvo-samsung-sdi-sign-up-wwf-call-temporary-ban-deep-sea-mining-2021-03-31/. For company specifics see: Microsoft. (2021). *Building a supply chain with integrity, accountability and respect* (Responsible Sourcing Report FY21). Microsoft Devices; ABN AMRO. (2021). *Exclusion List*; BBVA. (2020). *Environmental and Social Framework*; The Cooperative Bank. (2022). *Our customer-led Ethical Policy*; Hicks, R. (April 2022). Credit Suisse joins growing list of banks shunning deep-sea mining. EcoBusiness. https://www.eco-business.com/news/credit-suisse-joins-growing-list-of-banks-shunning-deep-sea-mining/; Standard Chartered. (2023). *Position Statement—Extractive industries*; Tridos Investment Management. (December 2021). *Tridos Bank supports moratorium on deep-sea mining*; and Patagonia. (December 2021). *Patagonia Calls for Moratorium on Seabed Mining*.

[3] For an overall discussion on companies against DSM see: WWF. (2021, December). *Business Backlash Against Deep-Sea Mining Grows*. https://wwf.panda.org/wwf_news/?4541466/Business-Backlash-Against-Deep-Sea-Mining-Grows; Greenpeace. (n.d.). *Race to the Top*. Stop Deep Sea Mining. Retrieved January 22, 2024, from https://www.racetothetop.app; Deep Sea Conservation Coalition. (n.d.). Voices calling for a moratorium: Companies and finance. Retrieved July 8, 2024, from https://deep-sea-conservation.org/solutions/no-deep-sea-mining/momentum-for-a-moratorium/companies-and-finance/; and Reid, H. (2021, March 31). Google, BMW, AB Volvo, Samsung back environmental call for pause on deep-sea mining. *Reuters*. https://www.reuters.com/

Fig. 1.6 Growing list of nations that have opposed DSM as of July 2025. (Source: Deep Sea Conservation Coalition, 2025. Reprinted with permission)

Other voices against DSM include more than 900 marine science and policy experts from more than 44 countries,[4] the European Parliament,[5] the Pacific Parliamentarians Alliance on Deep-Sea Mining,[6] the Parliament of the Canary Islands,[7] and the Regional Parliament of the Azores.[8] These voices are all echoing the sentiments of the 37 countries that have said no to DSM (Fig. 1.6).

Underwater cultural heritage has largely been left out of discussions about DSM's environmental impact while intangible cultural heritage has often been disregarded. There is a lack of baseline data to use as a benchmark to monitor DSM operations, and rules have yet to be developed for keeping scientists or archaeologists involved with their operations if they begin. As the archaeology chapter (Chap. 5) of this volume shows, telepresence and transmission of data is vital to studying deep-water environments and Ocean Heritage.

business/sustainable-business/google-bmw-volvo-samsung-sdi-sign-up-wwf-call-temporary-ban-deep-sea-mining-2021-03-31/. Ford has agreed to refrain from using deep sea minerals in the supply chain but has not announced support for a global moratorium (Greenpeace, *Race*).

[4] Deep Sea Conservation Coalition. (n.d.). Voices calling for a moratorium: Scientists. Retrieved July 8, 2024, from https://deep-sea-conservation.org/solutions/no-deep-sea-mining/momentum-for-a-moratorium/scientists/.

[5] Deep Sea Conservation Coalition. (n.d.). Voices calling for a moratorium: Governments and parliamentarians. Retrieved July 8, 2024, from https://deep-sea-conservation.org/solutions/no-deep-sea-mining/momentum-for-a-moratorium/governments-and-parliamentarians/.

[6] Ibid.

[7] Ibid.

[8] Ibid.

1.4 Steps Forward

The deep sea remains an area of the ocean in need of further study and scientists face large gaps in their understandings of its ecosystems let alone the UCH there. This makes it impossible to properly assess the potential impacts of DSM operations and as they stand, current ISA regulations are not sufficiently protective of Ocean Heritage. Global networks are hesitant to endorse the practice with the UN Environment Programme Finance Initiative (UNEP FI) concluding that financing DSM is inconsistent with principles of sustainable finance due to its significant reputational, regulatory, and operational risks. The International Capital Market Association (ICMA) issued guidance on sustainable blue economy bonds, explicitly excluding DSM. Similarly, the UN Conference on Trade and Development (UNCTAD) has excluded DSM from its sustainable ocean economy classification due to its high environmental risk.

Humans have a long history of detrimental activities in the ocean, but the deep sea represents 'one area where it is not too late and where we have a chance to get it right' (Cuyvers et al., 2018, p. vii). This point holds especially true when archaeologists consider the impacts of bottom trawling on their sites and the notable destruction this unregulated practice has caused (see Jarvis, 2024). With DSM, there is an opportunity to intervene before operations start. This was not successfully done with trawling and countless pieces of history have been lost. These mistakes cannot be repeated with DSM.

Tilot et al. (2021) explore current legal frameworks in the Pacific and 'an analysis of some traditional dimensions and cultural/societal aspects of marine resource management' to find that:

> the policies and practices developed in the Pacific could well serve as a suitable model elsewhere to reconcile commercial, ecological, cultural and social values within the context of deep sea mineral exploitation in addition to sustaining the Human Well-being and Sustainable Livelihoods (HWSL) of the Pacific communities and the health of the Global Ocean (p. 1, 2).

The authors note that many traditional marine management practices have flexible restrictions and can be applied in conjunction with other science-based tools such as Environmental Impact Assessments or permitting. Traditional knowledge and community-based marine management should be considered useful tools with a key role in this discussion (Tilot et al., 2021). These conversations traditionally have not found a place in the regulatory development at the ISA (Jarvis et al., 2023).

References

Amadi, E., & Mosnier, F. (2023). *The sky high cost of deep sea mining*. Planet Tracker. https://planet-tracker.org/wp-content/uploads/2023/06/Deep-Sea-Mining.pdf.

Amon, D. J., Gollner, S., Morato, T., Smith, C. R., Chen, C., Christiansen, S., Currie, B., Drazen, J. C., Fukushima, T., Gianni, M., Gjerde, K. M., Gooday, A. J., Grillo, G. G., Haeckel, M.,

Joyini, T., Ju, S.-J., Levin, L. A., Metaxas, A., Mianowicz, K., et al. (2022). Assessment of scientific gaps related to the effective environmental management of deep-seabed mining. *Marine Policy, 138*, 105006. https://doi.org/10.1016/j.marpol.2022.105006

Armstrong, C. W., Foley, N. S., Tinch, R., & Van Den Hove, S. (2012). Services from the deep: Steps towards valuation of deep sea goods and services. *Ecosystem Services, 2*, 2–13. https://doi.org/10.1016/j.ecoser.2012.07.001

Bernard, P. (2010). *Messages from the Deep: Water divinities, dreams and diviners in Southern Africa* [PhD]. Rhodes University.

Boschen, R. E., Rowden, A. A., Clark, M. R., Pallentin, A., & Gardner, J. P. A. (2016). Seafloor massive sulfide deposits support unique megafaunal assemblages: Implications for seabed mining and conservation. *Marine Environmental Research, 115*, 78–88. https://doi.org/10.1016/j.marenvres.2016.02.005

Brennan, M. L. (Ed.). (2024). *Threats to our ocean heritage: Potentially polluting wrecks*. Springer Nature Switzerland. https://doi.org/10.1007/978-3-031-57960-8

Costello, M. J., Cheung, A., & De Hauwere, N. (2010). Surface area and the seabed area, volume, depth, slope, and topographic variation for the world's seas, oceans, and countries. *Environmental Science & Technology, 44*(23), 8821–8828. https://doi.org/10.1021/es1012752

Cuyvers, L., Berry, W., Kristina, G., Torsten, T., & Caroline, W. (2018). *Deep seabed mining: A rising environmental challenge* (1st ed.). IUCN, International Union for Conservation of Nature. https://doi.org/10.2305/IUCN.CH.2018.16.en

Danovaro, R., Company, J. B., Corinaldesi, C., D'Onghia, G., Galil, B., Gambi, C., Gooday, A. J., Lampadariou, N., Luna, G. M., Morigi, C., Olu, K., Polymenakou, P., Ramirez-Llodra, E., Sabbatini, A., Sardà, F., Sibuet, M., & Tselepides, A. (2010). Deep-sea biodiversity in the Mediterranean sea: The known, the unknown, and the unknowable. *PLoS One, 5*(8), e11832. https://doi.org/10.1371/journal.pone.0011832

Danovaro, R., Fanelli, E., Aguzzi, J., Billett, D., Carugati, L., Corinaldesi, C., Dell'Anno, A., Gjerde, K., Jamieson, A. J., Kark, S., McClain, C., Levin, L., Levin, N., Ramirez-Llodra, E., Ruhl, H., Smith, C. R., Snelgrove, P. V. R., Thomsen, L., Van Dover, C. L., & Yasuhara, M. (2020). Ecological variables for developing a global deep-ocean monitoring and conservation strategy. *Nature Ecology & Evolution, 4*(2), 181–192. https://doi.org/10.1038/s41559-019-1091-z

De Jonge, D. S. W., Stratmann, T., Lins, L., Vanreusel, A., Purser, A., Marcon, Y., Rodrigues, C. F., Ravara, A., Esquete, P., Cunha, M. R., Simon-Lledó, E., Van Breugel, P., Sweetman, A. K., Soetaert, K., & Van Oevelen, D. (2020). Abyssal food-web model indicates faunal carbon flow recovery and impaired microbial loop 26 years after a sediment disturbance experiment. *Progress in Oceanography, 189*, 102446. https://doi.org/10.1016/j.pocean.2020.102446

Dobush, B.-J., & Warner, M. (2024). *Deep sea mining isn't worth the risk: High costs, financial developments since 2021, and externalities stand to diminish theoretical returns on investments*. The Ocean Foundation. https://oceanfdn.org/wp-content/uploads/2024/02/dsm-finance-brief-2024.pdf.

DOSI. (2022). *What does the deep ocean do for you?* [Policy Brief]. Deep-Ocean Stewardship Initiative. https://www.dosi-project.org/wp-content/uploads/deep-ocean-ecosystem-services-brief.pdf.

DOSI. (2023a). *Polymetallic nodule-rich Abyssal plains* [Information Sheet]. Deep-Ocean Stewardship Initiative. https://www.dosi-project.org/wp-content/uploads/abyssal-plain-info-sheet.pdf.

DOSI. (2023b). *The seamount ecosystem* [Information Sheet]. Deep-Ocean Stewardship Initiative. https://www.dosi-project.org/wp-content/uploads/seamount-info-sheet.pdf.

DOSI. (2023c). *Hydrothermal vent ecosystems* [Information Sheet]. Deep-Ocean Stewardship Initiative. https://www.dosi-project.org/wp-content/uploads/vents-ecosystems-info-sheet.pdf.

Drazen, J. C., Smith, C. R., Gjerde, K. M., Haddock, S. H. D., Carter, G. S., Choy, C. A., Clark, M. R., Dutrieux, P., Goetze, E., Hauton, C., Hatta, M., Koslow, J. A., Leitner, A. B., Pacini, A., Perelman, J. N., Peacock, T., Sutton, T. T., Watling, L., & Yamamoto, H. (2020). Midwater ecosystems must be considered when evaluating environmental risks of deep-sea mining. *Proceedings of the National Academy of Sciences, 117*(30), 17455–17460. https://doi.org/10.1073/pnas.2011914117

Erickson, K. L., Macko, S. A., & Van Dover, C. L. (2009). Evidence for a chemoautotrophically based food web at inactive hydrothermal vents (Manus Basin). *Deep Sea Research Part II: Topical Studies in Oceanography, 56*(19–20), 1577–1585. https://doi.org/10.1016/j.dsr2.2009.05.002

European Academics Science Advisory Council (EASAC). (2023). *Deep-Sea Mining: Assessing evidence on future needs and environmental impacts*. https://easac.eu/fileadmin/user_upload/EASAC_Deep_Sea_Mining_Web_publication_.pdf.

Everett, J., Kammen, D., & Rowland, S. (2023). *Next generation EV batteries eliminate the need for deep sea mining* [White Paper]. Blue Climate Initiative.

Fedje, D. W., & Mathewes, R. W. (2005). *Haida Gwaii: Human history and environment from the time of loon to the time of the iron people*. University of British Columbia Press.

Filho L. W., Abubakar, I., Nunes, C., Platje, J., Ozuyar, P., Will, M., Nagy, G., Al-Amin, A., Hunt, J., & Li, C. (2021). Deep Seabed Mining: A Note on Some Potentials and Risks to the Sustainable Mineral Extraction from the Oceans. *Journal of Marine Science and Engineering, 9(5), 521*. https://doi.org/10.3390/jmse9050521.

Gianni, M. (2004). *High seas bottom trawl fisheries and their impacts on the biodiversity of vulnerable deep-sea ecosystems*. IUCN/the World Conservation Union; Natural Resources Defense Council; WWF International; Conservation International. https://portals.iucn.org/library/sites/library/files/documents/Rep-2004-001.pdf

Gruber, N., Clement, D., Carter, B. R., Feely, R. A., Van Heuven, S., Hoppema, M., Ishii, M., Key, R. M., Kozyr, A., Lauvset, S. K., Lo Monaco, C., Mathis, J. T., Murata, A., Olsen, A., Perez, F. F., Sabine, C. L., Tanhua, T., & Wanninkhof, R. (2019). The oceanic sink for anthropogenic CO_2 from 1994 to 2007. *Science, 363*(6432), 1193–1199. https://doi.org/10.1126/science.aau5153

Hamley, G. J. (2022). The implications of seabed mining in the Area for the human right to health. *Review of European, Comparative & International Environmental Law, 31*(3), 389–398. https://doi.org/10.1111/reel.12471

Hilmi, N., Sutherland, M., Farahmand, S., Haraldsson, G., Van Doorn, E., Ernst, E., Wisz, M. S., Claudel Rusin, A., Elsler, L. G., & Levin, L. A. (2023). Deep sea nature-based solutions to climate change. *Frontiers in Climate, 5*, 1169665. https://doi.org/10.3389/fclim.2023.1169665

Impossible Metals. (2024). Robotic collection system. *Impossible Metals*. https://impossiblemetals.com/technology/robotic-collection-system/.

IUCN. (2022). *Deep-Sea mining* [Issues Brief]. International Union for Conservation of Nature and Natural Resources. https://www.iucn.org/sites/default/files/2022-07/iucn-issues-brief_dsm_update_final.pdf.

Jaeckel, A., Harden-Davies, H., Amon, D. J., Van Der Grient, J., Hanich, Q., Van Leeuwen, J., Niner, H. J., & Seto, K. (2023). Deep seabed mining lacks social legitimacy. *Npj Ocean Sustainability, 2*(1), 1. https://doi.org/10.1038/s44183-023-00009-7

Jarvis, C. (Ed.). (2024). *Threats to our ocean heritage: Bottom trawling*. Springer Nature Switzerland. https://doi.org/10.1007/978-3-031-57953-0

Jarvis, C., Ermida, M. P., & Varmer, O. (2023). Threats to underwater cultural heritage from existing and future human activities. *Blue Papers, 2*(1), 76–83. https://doi.org/10.58981/bluepapers.2023.1.08

Jeffery, B., McKinnon, J. F., & Van Tilburg, H. (2021). Underwater cultural heritage in the Pacific: Themes and future directions. *International Journal of Asia Pacific Studies, 17*(2), 135–168. https://doi.org/10.21315/ijaps2021.17.2.6

Jenner, N., Benbow, S., Weller, C., & Howard, P. (2023). *Update to 'An assessment of the risks and impacts of seabed mining on marine ecosystems': We now know more, but still not enough to rush into deep sea mining*. Fauna & Flora. https://www.fauna-flora.org/wp-content/uploads/2023/05/fauna-flora-deep-sea-mining-update-report-march-23.pdf.

Jones, D. O. B., Amon, D. J., & Chapman, A. S. A. (2018). Mining deep-ocean mineral deposits: What are the ecological risks? *Elements, 14*, 325–330. https://doi.org/10.2138/gselements.14.5.325

Kröger, K. (2024, September 5). *Deep sea mining: The mineral resources, the environmental impacts and the scientific gaps*. IUCN Webinar. CEM Dialogue: Deep Sea Mining and what is expected as we continue to uphold IUCN Resolution 122. https://youtu.be/_vYUA%2D%2DFNfU?si=vvkt-i6IUrn5H6Jy.

Le, J. T., Levin, L. A., & Carson, R. T. (2017). Incorporating ecosystem services into environmental management of deep-seabed mining. *Deep Sea Research Part II: Topical Studies in Oceanography, 137*, 486–503. https://doi.org/10.1016/j.dsr2.2016.08.007

Levin, L. A., Mengerink, K., Gjerde, K. M., Rowden, A. A., Van Dover, C. L., Clark, M. R., Ramirez-Llodra, E., Currie, B., Smith, C. R., Sato, K. N., Gallo, N., Sweetman, A. K., Lily, H., Armstrong, C. W., & Brider, J. (2016). Defining "serious harm" to the marine environment in the context of deep-seabed mining. *Marine Policy, 74*, 245–259. https://doi.org/10.1016/j.marpol.2016.09.032

Miller, K. A., Brigden, K., Santillo, D., Currie, D., Johnston, P., & Thompson, K. F. (2021). Challenging the need for deep seabed mining from the perspective of metal demand, biodiversity, ecosystems services, and benefit sharing. *Frontiers in Marine Science, 8*, 706161. https://doi.org/10.3389/fmars.2021.706161

Miller, K. A., Thompson, K. F., Johnston, P., & Santillo, D. (2018). An overview of seabed mining including the current state of development, environmental impacts, and knowledge gaps. *Frontiers in Marine Science, 4*, 418. https://doi.org/10.3389/fmars.2017.00418

Muñoz-Royo, C., Peacock, T., Alford, M. H., Smith, J. A., Le Boyer, A., Kulkarni, C. S., Lermusiaux, P. F. J., Haley, P. J., Mirabito, C., Wang, D., Adams, E. E., Ouillon, R., Breugem, A., Decrop, B., Lanckriet, T., Supekar, R. B., Rzeznik, A. J., Gartman, A., & Ju, S.-J. (2021). Extent of impact of deep-sea nodule mining midwater plumes is influenced by sediment loading, turbulence and thresholds. *Communications Earth & Environment, 2*(1), 148. https://doi.org/10.1038/s43247-021-00213-8

NOAA Ocean Exploration. (n.d.). *Hydrothermal vents* [Fact Sheet]. https://oceanexplorer.noaa.gov/edu/materials/hydrothermal-vents-fact-sheet.pdf.

Pickens, C., Lily, H., Harrould-Kolieb, E., Blanchard, C., & Chakraborty, A. (2024). From what-if to what-now: Status of the deep-sea mining regulations and underlying drivers for outstanding issues. *Marine Policy, 169*, 105967. https://doi.org/10.1016/j.marpol.2023.105967

Rabone, M., Wiethase, J. H., Simon-Lledó, E., Emery, A. M., Jones, D. O. B., Dahlgren, T. G., Bribiesca-Contreras, G., Wiklund, H., Horton, T., & Glover, A. G. (2023). How many metazoan species live in the world's largest mineral exploration region? *Current Biology, 33*(12), 2383–2396.e5. https://doi.org/10.1016/j.cub.2023.04.052

Ramirez-Llodra, E. (2020). Deep-sea ecosystems: Biodiversity and anthropogenic impacts. In C. Banet (Ed.), *The law of the seabed* (pp. 36–60). Brill | Nijhoff. https://doi.org/10.1163/9789004391567_004

Rogers, A. D. (2018). The biology of seamounts: 25 Years on. In *Advances in marine biology* (Vol. 79, pp. 137–224). Elsevier. https://doi.org/10.1016/bs.amb.2018.06.001

Simon-Lledó, E., Bett, B. J., Huvenne, V. A. I., Köser, K., Schoening, T., Greinert, J., & Jones, D. O. B. (2019). Biological effects 26 years after simulated deep-sea mining. *Scientific Reports, 9*(1), 8040. https://doi.org/10.1038/s41598-019-44492-w

Singh, P. A. (2021). The two-year deadline to complete the International Seabed Authority's Mining Code: Key outstanding matters that still need to be resolved. *Marine Policy, 134, 104804*. https://doi.org/10.1016/j.marpol.2021.104804.

Singh, P. A., Jaeckel, A., & Ardron, J. A. (2025). A pause or moratorium for deep seabed mining in the area? The legal basis, potential pathways, and possible policy implications. *Ocean Development & International Law, 56*(1), 18–44. https://doi.org/10.1080/00908320.2024.2439877

Smith, C. R., Tunnicliffe, V., Colaço, A., Drazen, J. C., Gollner, S., Levin, L. A., Mestre, N. C., Metaxas, A., Molodtsova, T. N., Morato, T., Sweetman, A. K., Washburn, T., & Amon, D. J. (2020). Deep-sea misconceptions cause underestimation of seabed-mining impacts. *Trends in Ecology & Evolution, 35*(10), 853–857. https://doi.org/10.1016/j.tree.2020.07.002

Stratmann, T., Soetaert, K., Kersken, D., & Van Oevelen, D. (2021). Polymetallic nodules are essential for food-web integrity of a prospective deep-seabed mining area in Pacific abyssal plains. *Scientific Reports, 11*(1), 12238. https://doi.org/10.1038/s41598-021-91703-4

Sweetman, A. K., Smith, A. J., De Jonge, D. S. W., Hahn, T., Schroedl, P., Silverstein, M., Andrade, C., Edwards, R. L., Lough, A. J. M., Woulds, C., Homoky, W. B., Koschinsky, A., Fuchs, S., Kuhn, T., Geiger, F., & Marlow, J. J. (2024). Evidence of dark oxygen production at the abyssal seafloor. *Nature Geoscience, 17*(8), 737–739. https://doi.org/10.1038/s41561-024-01480-8

The Environmental Justice Foundation. (2023). *Towards the Abyss: How the rush to deep-sea mining threatens people and our planet*. The Environmental Justice Foundation. https://ejfoundation.org/resources/downloads/towards-the-abyss-ejf-deep-sea-mining-report.pdf.

The Metals Company. (2023). *Annual report 2022*.

The Ocean Foundation. (2023). *Deep seabed mining* [Research Page]. https://oceanfdn.org/deep-seabed-mining/.

Thurber, A. R., Sweetman, A. K., Narayanaswamy, B. E., Jones, D. O. B., Ingels, J., & Hansman, R. L. (2014). Ecosystem function and services provided by the deep sea. *Biogeosciences, 11*(14), 3941–3963. https://doi.org/10.5194/bg-11-3941-2014

Tilot, V., Willaert, K., Guilloux, B., Chen, W., Mulalap, C. Y., Gaulme, F., Bambridge, T., Peters, K., & Dahl, A. (2021). Traditional dimensions of seabed resource management in the context of deep sea mining in the Pacific: Learning from the socio-ecological interconnectivity between island communities and the ocean realm. *Frontiers in Marine Science, 8*, 637938. https://doi.org/10.3389/fmars.2021.637938

Turner, P. J., Cannon, S., DeLand, S., Delgado, J. P., Eltis, D., Halpin, P. N., Kanu, M. I., Sussman, C. S., Varmer, O., & Van Dover, C. L. (2020). Memorializing the middle passage on the Atlantic seabed in areas beyond national jurisdiction. *Marine Policy, 122*, 104254. https://doi.org/10.1016/j.marpol.2020.104254

United Nations Environment Programme Finance Initiative. (2022). *Harmful marine extractives: Understanding the risks & impacts of financing non-renewable extractive industries*. https://www.unepfi.org/wordpress/wp-content/uploads/2022/05/Harmful-Marine-Extractives-Deep-Sea-Mining.pdf.

Van Dover, C. L., Grassle, J. F., & Boudrias, M. (1990). Hydrothermal Vent Fauna of Escanaba Trough (Gorda Ridge). In G. R. McMurray (Ed.), *Gorda Ridge* (pp. 285–287). Springer. https://doi.org/10.1007/978-1-4612-3258-2_21

Van Dover, C. L., Arnaud-Haond, S., Gianni, M., Helmreich, S., Huber, J. A., Jaeckel, A. L., Metaxas, A., Pendleton, L. H., Petersen, S., Ramirez-Llodra, E., Steinberg, P. E., Tunnicliffe, V., & Yamamoto, H. (2018). Scientific rationale and international obligations for protection of active hydrothermal vent ecosystems from deep-sea mining. *Marine Policy, 90*, 20–28. https://doi.org/10.1016/j.marpol.2018.01.020

Watson, A. J., Schuster, U., Shutler, J. D., Holding, T., Ashton, I. G. C., Landschützer, P., Woolf, D. K., & Goddijn-Murphy, L. (2020). Revised estimates of ocean-atmosphere CO2 flux are consistent with ocean carbon inventory. *Nature Communications, 11*(1), 4422. https://doi.org/10.1038/s41467-020-18203-3

Weaver, P. P. E., Billett, D. S. M., & van Dover, C. L. (2018). Environmental risks of deep-sea mining. In M. Salomon & T. Markus (Eds.), *Handbook on marine environment protection: Science, impacts and sustainable management* (Vol. 1, pp. 215–246). Springer.

Yeh, W., & The World Economic Forum. (2022). *Decision-making on deep-sea mineral stewardship: A supply chain perspective*. The World Economic Forum. https://www3.weforum.org/docs/WEF_Decision_Making_on_Deep_Sea_Mineral_Stewardship_2022.pdf.

Yesson, C., Clark, M. R., Taylor, M. L., & Rogers, A. D. (2011). The global distribution of seamounts based on 30 arc seconds bathymetry data. *Deep Sea Research Part I: Oceanographic Research Papers, 58(4), 442–453*. https://doi.org/10.1016/j.dsr.2011.02.004.

Chapter 2
Deep Sea Mining vs. Underwater Cultural Heritage

Maria Pena Ermida

Abstract This chapter seeks to provide an overview of the legal framework surrounding the protection of UCH as a part of the Marine Environment within the context of Law of the Sea, focusing particularly on the rules regarding deep sea mining. The text seeks to analyse how such an activity may be considered compatible, or not with the protection of the marine environment and in particular how underwater cultural heritage may be addressed within such a context.

Keywords Marine environment · Area · Mining · Law of the sea · Underwater cultural heritage

2.1 Introduction

As Humanity comes closer to exploring the bottom of the ocean, questions concerning our knowledge of this area emerge. Usually, such questions regard the environmental aspect of the deep sea and impacts on marine organisms and ecosystems. However, a less common question is, how will such an activity impact Humanity's shared history, which still lies hidden at the bottom of the sea? How can one protect underwater cultural heritage (UCH) using the legal framework from an activity such as deep sea mining (DSM)?

The framework concerning the protection of UCH results from the interaction between different legal frameworks, namely the law of the sea, environmental law, and cultural heritage. Various legal instruments will be referred to, and although

---

M. P. Ermida (✉)
Católica Research Centre for the Future of Law, Faculty of Law – Universidade Católica Portuguesa, Lisbon, Portugal
e-mail: mermida@ucp.pt

C. Jarvis (ed.), *Threats to Our Ocean Heritage: Deep Sea Mining*,
SpringerBriefs in Archaeology, https://doi.org/10.1007/978-3-031-98238-5_2

they may seem unrelated at first, the reader will be guided through their unifying points concerning the protection of UCH as a part of the marine environment, considering the duty to protect UCH and the relevance of the application of the precautionary principle.

2.2 From Total Freedom to Common Heritage of Mankind[1]

Before the adoption of the 1982 Law of the Sea Convention (LOSC),[2] there was no explicit agreement on how to share, or exploit, minerals found on the seabed. Although there were some possible methods, the practical result of any of those theories would be a 'first come, first served' scenario, whereby developed States with the technology to carry out such endeavours would be better placed to exploit resources in the terms and timeframe they saw fit (Tanaka, 2019, p. 218). At the same time, this activity had the potential to negatively impact States that relied on the export of land-based mining for their economy, many of which were developing States.[3] A solution was needed to avoid difficulties which would deepen the existing rift between developed and developing States further.

To give States a purpose to come together amid Cold War uncertainty and perceived unfairness, Maltese Ambassador Arvid Pardo made a historic speech in 1967 at the UN General Assembly, stating that a new framework of cooperation amongst States was vital to prevent the unruled appropriation by States of the deep sea for military purposes, the exploitation of marine resources in an unequal manner, and the dumping of radioactive waste in these areas (Pardo, 1967). In light of these risks, Pardo proposed a moratorium on all uses of the deep sea until the adoption of a legal framework that put these areas and resources under the Principle of Common Heritage of Mankind (CHM).[4] This would eventually happen with the 1969 Moratorium, General Assembly Resolution 2574 D (XXIV), known as the Moratorium Resolution. According to it, no activities targeted at exploiting resources were to occur in the deep sea in areas beyond national jurisdiction until an international framework was adopted for this area and the resources therein.

---

[1] In the twenty-first century, this concept is most often referred to as the Common Heritage of Humankind, but the acronym CHM is still in use.

[2] The LOSC, also referred to as the Constitution of the Oceans, is the result of more than a decade of negotiations, political declarations, workgroups, and heated discussions that culminated in the adoption of one of the most comprehensive and sophisticated legal instruments in the history of international law.

[3] Take for example the Democratic Republic of the Congo, rich in minerals such as copper, cobalt, diamonds, and gold, or Bolivia rich in tin, silver, and zinc.

[4] Although Arvid Pardo did not conceptualize this Principle, he was the father of its introduction in the Law of the Sea. This Principle had been included in other conventions such as the *Antarctic Treaty (1961),* the Preamble of the NATO treaty (1949), and the Treaty on the Non-Proliferation of Nuclear Weapons (1968).

Pardo (1975) advocated that to give relevance to values such as freedom or sovereignty instead of a principle like Common Heritage of Humankind, would result in 'a competitive scramble for sovereign rights over the land underlying the world's seas and oceans, surpassing in magnitude and its implications last century's colonial scramble for territory in Asia and Africa' (Pardo, 1975, p. 31). This would raise the same suspicions and tensions among the powerful few, something Pardo was attempting to avoid, and further the resentment of former colonies towards their colonisers. The answer was, therefore, obvious—to develop an approach based on the prohibition of national appropriation, peaceful uses, and a shared and responsible utilisation of the Area, with consideration for developing countries. The goal was to benefit Humankind instead of the same established few (Vönecky & Höfelmeier, 2017, mn 9).

Ultimately, this principle made it into the LOSC. This Convention divided the ocean into Areas Within National Jurisdiction and Areas Beyond National Jurisdiction. Areas Within the National Jurisdiction of States include the Territorial Sea (LOSC 1982, Article 2), the Contiguous Zone (LOSC 1982, Article 33), the Exclusive Economic Zone (EEZ), and the Continental Shelf (LOSC 1982, Article 76). Areas beyond National Jurisdiction include the High Seas (the water column and resources therein beyond the EEZ of States) and the Area (the seabed beyond the continental shelves of States) (LOSC 1982, Article 1 (1) (1), Article 133 *et seq.*; LOSC 1982, Part XI). As a widely accepted international agreement ratified by 169 nations plus the European Union, most of LOSC is recognised as codified customary international law. As a result, over the past four decades, it has been followed in practice by parties and non-parties, with many rights and obligations resulting from this treaty being pointed at as customary international law (Churchill, 2022, 32 *et seq.*) (Fig. 2.1).

With the adoption of the LOSC, the Area and its resources became governed by the Principle of the CHM which had three elements. (LOSC, 1982, Article 136; on this Principle see Wolfrum, 1983; Kiss, 1985; Pureza, 1998; Scovazzi, 2004) Firstly, the creation of an international management structure for the Area and its resources that would guarantee that all benefits of activities in the area were shared equally among Humanity (LOSC, 1982, Articles 136, 137, 138 & 140). Secondly, the use of the Area exclusively for peaceful purposes (LOSC, 1982, Article 141). Finally, the principle of non-appropriation by States and consequential benefit sharing of the resources therein (LOSC, 1982, Article 137). This chapter explores how this concept is relevant to UCH.

The inclusion of 'The Area and its resources' in the principle of the CHM has raised much discussion concerning the nature of the benefits to be gained from the exploitation of resources (Vönecky & Höfelmeier, 2017, mn 22).[5] For this text, it is essential to analyse the evolution of the Principle of the CHM, namely in terms of

[5] By ensuring technology and knowledge transfer, Articles 143 and 144 of the LOSC ensures the sharing of benefits other than economic among all States, including developing States. This framework has however suffered some changes through the Implementation Agreement for Part XI of the Convention.

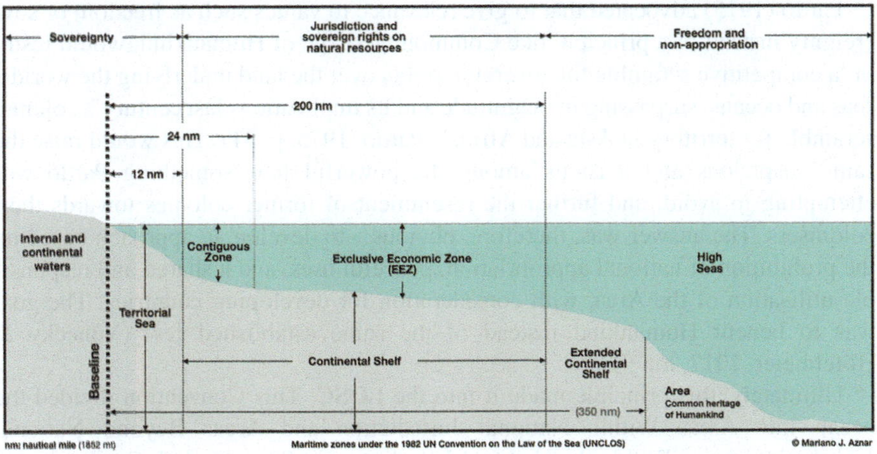

Fig. 2.1 The maritime zones include the 12 nm Territorial Sea, the 24 nm Contiguous Zone, the 200 nm Exclusive Economic Zone, and the High Seas. (Source: © Mariano Aznar, reprinted with permission)

its benefit-sharing aspects (LOSC, 1982, Article 137 & 140). Initially, this element included an idea of cooperation regarding these resources that mainly targeted the economic benefit one could extract from them. This means that the Principle of the CHM could also be applied, particularly its benefit-sharing element, to other activities, such as activities with a research or knowledge-seeking character (Vönecky & Höfelmeier, 2017, mn 22; See also Scovazzi, 2007; Heafy, 2014). Secondly, the CHM principle has also been understood to have both a spatial dimension (meaning that it applies to all states (AO ITLOS 2011, para. [159]) and a temporal dimension (allowing it to consider both present and future generations) (Vönecky & Höfelmeier, 2017, mn 23; See also LOSC, 1982, Articles 137, 138, & 140), thus raising the question of Intergenerational Justice. In this sense, it could be argued that cultural benefits, the knowledge of our human history, and its passing to future generations is something that figures within the scope of the principle of the CHM and thus a duty to protect UCH may naturally arise. This position, as shall be demonstrated, can be found in the text of the LOSC, namely through Articles 303 and 149. However, before analysing these provisions it is important to provide a general overview of the framework for deep sea mining according to the LOSC.

2.3 Deep Sea Mining Beyond the Law

The Area includes 'the seabed and ocean floor and subsoil thereof, beyond the limits of national jurisdiction' (LOSC, 1982, Article 1(1) (1)) and is addressed by LOSC, Part XI. The administration of the Area and its resources is carried out by the International Seabed Authority (ISA), one of the institutions created by LOSC (along with ITLOS and the Commission on the Limits of the Continental Shelf). This institution's mandate includes regulating the 'activities in the Area', which include 'drilling, dredging, excavation, disposal of waste, construction and operation or maintenance of installations, pipelines and other devices related to such activities' (LOSC, 1982, Article 145 (a) and Annex III, Article 17 (2) (f))). No member State may authorise activities in the Area without the knowledge and permission of the Authority.[6] If an entity wishes to pursue any activity in the Area, it must submit plans of work to the Authority for approval as stated in Article 3 (1) of Annex III of LOSC. According to Article 153 (2), these entities must be sponsored by that State.

The role of the Sponsoring State is to contribute to the common interest of all States in the proper implementation of the principle of the Common Heritage of Humankind by assisting the Authority and by acting on its own to ensure that entities under its jurisdiction conform to the rules on deep sea mining (AO ITLOS 2011, para. [226]).

According to Article 139 (1) of the LOSC 'States Parties shall be responsible for ensuring that activities in the Area (…) shall be carried out in conformity with this Part. (…).'. Annex III, Article 4 (4) reinforces this, stating that 'The sponsoring State or States shall, pursuant to Article 139, have the responsibility to ensure, within their legal systems, that a contractor so sponsored shall carry out activities in the Area in conformity with the terms of its contract and its obligations under this Convention. A sponsoring State shall not, however, be liable for damage caused by any failure of a contractor sponsored by it to comply with its obligations if that State Party has adopted laws and regulations and taken administrative measures that are, within the framework of its legal system, reasonably appropriate for securing compliance by persons under its jurisdiction'.

In the last 40 years, the international community has walked towards regulating activities with a hazardous character based on a logic of 'community interest' or 'common concern' (AO ITLOS 2011, para. [76]). In addition, it has also been established that the obligation of the Sponsoring State, based on both provisions, was one of due diligence, meaning that the State is required to adopt a specific conduct or refrain from doing so. In this case, States were called upon to apply 'reasonably appropriate' measures, legal and administrative, to ensure compliance with the relevant rules and thus prevent harm and violations (Ago, 1977, p. 8; See also Zhang, 2013). This has created much space for ambiguity in the adopted measures.

With time, scholarship and case law have attempted to densify the term 'reasonably appropriate'. On the one hand there would need to be a domestic legal framework that included measures considered 'reasonably appropriate' (See *Pulp Mills*,

[6] The question of non-member States will not be addressed in detail in this chapter. However, it is important to note that although much of the law of the Sea is considered to be customary law, this does not fully encompass the framework for the Area and its resources. Even so, there is some room to argue that some non-members States have taken actions that counteract this tendency, such as US's signing of the 1994 Agreement of part XI of the LOSC. For more on this issue see Vönecky, S. and Höfelmeier, C. (2017), mn 8–9.

2010, para. [187]; AO ITLOS 2011[123], [215] and [227]). Moreover, the direct obligations specific to the quality of the Sponsoring State, meaning are those with which the Sponsor States must comply 'independently of their obligation to ensure a certain behavior by the sponsored contractor', should also be taken into account, (AO ITLOS 2011[121]) namely the obligation to assist the Authority, (AO ITLOS 2011[124]; LOSC, 1982, Article 153(4) *in fine*) the obligation to apply best environmental practices(AO ITLOS 2011[135-136]; ISA Regulation, 2009, article 33 (2)), the obligation to protect the marine environment, (AO ITLOS 2011[138]; LOSC, 1982, Article 209) and the application of the precautionary approach (AO ITLOS 2011 [125-135]; ISA Regulation, 2013, Regulation 31(2)). Finally, the adopted domestic measures should be no 'less effective than the international rules, regulations and procedures' to be adopted by the ISA (LOSC, 1982, Articles 145 & 209(2)). However, The ISA has been heavily criticised for not including adequate environmental and sustainability standards in such instruments (ISA, 2020). Presently, the Council of ISA is still negotiating the so-called exploitation regulations. At this point, there is still no verdict regarding what these regulations are bound to look like (See Blanchard et al., 2023). A moratorium until all of the concerns for the marine environment and related cultural heritage would be a more prudent precautionary approach.

Nevertheless, it is worth mentioning that, unlike in the past, those who seek the protection of UCH have been called to speak concerning its relationship with an activity such as deep sea mining. For instance, recently, there has been ongoing pressure by NGOs to include that identifies the UCH that members of the ISA have a duty to protect under LOSC Articles 149, 303(1), and Part XII in the baseline surveys of the marine environment for environmental impact assessments, as a way of ensuring science-based decision making when granting licenses to conduct mining.[7] Moreover, UCH and natural heritage are physically and legally linked. Thus, integrating the legal obligations to protect UCH and the marine environment into Environmental Impact Assessments and consideration as areas to set aside from exploitation or as MPAs consistent with the BBNJ are crucial for their adequate protection.

2.4 The Treatment of UCH in the LOSC

Before addressing the treatment of UCH in the LOSC, it is important to note that there were already a number of instruments relevant for the establishment of a general duty to protect UCH prior to the LOSC. Take for example, the 1972 Stockholm Declaration which is considered to have been instrumental in the 'development of

[7]This is in line with recent State Practice. In fact, the EU Directive on Environmental Impacts amendments in 2014 referred to the need to assess the effects of projects on, biodiversity, water and even 'cultural heritage, including architectural and archaeological aspects....' (Annex IV – Information for the Environmental Impact Assessment Report).

provisions that resulted in the 1982 Law of the Sea Convention and the framework for the conduct of activities at sea, which must consider the duties to protect our natural and cultural heritage' (Varmer, 2020, p. 88). In addition, in 1972 the World Heritage Convention also took place, playing a significant role in the establishment of a duty to protect UCH and influencing later calls for recognition of cultural heritage in the high seas, including wreck sites such as *Titanic* (Varmer, 2020, p. 95; Ermida, 2024, p. 29). These shall be addressed in more detail further in this text.

It is, however, undeniable that during the LOSC negotiations, UCH was not an issue that received much attention (Risvas, 2013, p. 564; Scovazzi, 2017b, mn. 16). While some States strongly supported the regulation of such matters within an instrument such as the LOSC, others were firmly against it (Risvas, 2013, p. 565). Despite this stalemate, the drafters of the LOSC managed to include the 'protection' and 'preservation' of our cultural heritage in the final text, namely in Articles 149 and 303. Although they referred to UCH as 'archaeological and historical objects',[8] one could argue that although both provisions express a 'duty to protect' these objects, the provisions are fairly general and vague for implanting the duty to protect UCH (Scovazzi, 2017a, mn. 9; Ermida, 2024, p. 30). Moreover, it is also essential to note that while Article 303 is in the general provisions of Part XVI of the LOSC, Article 149 is included in Part XI of the LOSC, thus regarding the Area specifically. As we shall see, this holds specific implications for UCH, particularly in cases where DSM conflicts with the duty to protect the marine environment present in the LOSC, namely in Articles 145 and 192 (Czybulka, 2017, mn. 25; Varmer, 2020, p. 92; Ermida, 2024, p. 32).

Article 303 seeks to establish two general duties: the duty to protect UCH and the duty to cooperate in doing so (Scovazzi, 2017b, mn. 10). These duties apply to our heritage in all maritime zones, including the Area. However, a closer analysis of this provision shows that that they are notably vague (Scovazzi, 2017b, mn 10; See also Aznar, 2022). For instance, article 303 (2) limits the geographic scope of coastal State jurisdiction to control and regulate activities directed at UCH. There are no doubts regarding the competence of the Coastal State regarding the territorial sea where, according to Article 2 of the LOSC, the State benefits from total sovereignty, with only a few limitations such as the right of innocent passage, present in Article 17 of the LOSC. However, such doubts arise when activities in the contiguous zone come to play, particularly between the 12 and the 24 nautical-miles limit. The reading of Article 303(2) suggests that the removal of archaeological or historical objects located within this limit could entail a violation of legislation of the Coastal State, despite it saying nothing about cultural heritage. Thus, it could be argued that the text and logic of the LOC suggests that UCH should not be protected *per se* but that it should be protected incidentally when the coastal State is entitled to exercise

[8] A seemingly deceiving provision which should however be read in a broad sense to include artifacts undoubtedly within the field of archaeology but also those of relatively recent origin but that hold a historical weight, such as a sunken ship from WW2. For this reason, throughout this text, for a question of clarity the term used to describe such artifacts will be considered UCH as defined in the 2001 Convention on Underwater Cultural Heritage.

jurisdiction due to issues like smuggling, public health or immigration (Scovazzi, 2017b, mn. 13). Moreover, the wording of the provision itself does not even predict *in situ* protection rendering coastal States powerless in scenarios where UCH is at risk of being destroyed in the very place where it stands due to economic activities such as deep sea mining (Scovazzi, 2017b, mn. 14). Luckily, recently the ICJ has provided important insights regarding the treatment of UCH within the Contiguous Zone, stating that Article 303 (2) reflects customary international law and that States may extend the application of their own cultural heritage legislation over the contiguous zone 'in so far as it includes the power of control with respect to archaeological and historical objects found within the contiguous zone'. See Alleged Violations of Sovereign Rights and Maritime Spaces in the Caribbean Sea (Nicaragua v. Colombia), Judgment, I.C.J. Reports 2022, p. 266, [180–185]).

It is also worth noting that the LOSC defines nothing regarding archaeological and historical objects found on the Continental shelf or the EEZ. In the CS, for instance, in accordance with Article 77 (1), the coastal State only holds rights to explore and exploit 'natural resources'—a concept that does not easily encompass UCH (Scovazzi, 2017b, mn. 19; See also Oxman, 1988; Aznar, 2014). This leaves a legal vacuum behind begging for the Principle of freedom of the high seas to intervene (Scovazzi, 2017b, mn. 20).[9] There are, however, ways around this. For instance, in accordance with Articles 33, 56 and 94 of the LOSC a coastal State has jurisdiction over its nationals and vessels and a State may exercise jurisdiction over vessels flying its flag in its EEZ/CS and on the high seas.[10] However, if what is at stake is a foreign national or vessel conducting a highly hazardous activity or even a case of UCH which is a clear part of the marine environment, it could be argued that this could trigger jurisdiction over natural resources in the EEZ/CS from UCH's treasure hunters, salvors, or looters, a coastal State may enforce its natural resource regulations against those looters and salvors. Another example lies in the coastal States jurisdiction, authority, and control over the placement and management of artificial reefs. It could be argued that as UCH can also serve as an artificial reef, a coastal State could have ample authority to protect, manage, and prevent looting and unwanted salvage under that regime. Nevertheless, although the text in Article 303(4) implicitly recognises that there is little guidance or detail concerning the scope and reach of this duty, it contemplates a subsequent agreement (Scovazzi, 2017b mn. 42; See also Aznar, 2022).

Article 149, on the other hand, makes an apparent reference to activities concerning the non-living resources of the deep sea in the Area. According to Article 149, 'all objects of an archaeological and historical nature found in the Area shall be

[9] Predicting that the details needed for this new area of international underwater heritage law would be addressed by State practice and presumably in a new UNESCO Convention.

[10] This is why in the Titanic agreement the US and UK use that flag State jurisdiction to protect *Titanic,* which lies on the outer continental shelf of Canada under the high seas. This is in line with the LOSC drafters' intention to avoid any significant erosion of the principle of freedom of the high seas, particularly in regard to the 'creeping jurisdiction' of coastal States in areas beyond the territorial sea (Aznar & Varmer, 2013).

preserved or disposed of for the benefit of mankind as a whole, particular regard being paid to the preferential rights of the State or country of origin, or the State of cultural origin, or the State of historical and archaeological origin'. From this provision, one may extract that all members of the ISA have an obligation to address this duty to protect UCH, namely through the regulations the ISA must adopt for deep sea mining. This also shows that the drafters preferred to protect UCH for the common good over private interest. Considering the language used ('for the benefit of mankind as a whole'), such objects in the Area appear to have some relationship to the protection to the Area and its natural resources, which are subject to the principle of the CHM. One could even argue that UCH is also under the principle of CHM (in this sense, see Scovazzi, 2017a).

Finally, one cannot ignore that Article 192 of the LOSC establishes a 'general obligation to protect and preserve the marine environment' (LOSC, 1982, article 192). Although the concept of 'marine environment' is not defined in Article 1 of the LOSC, it could be argued, taking from the preparatory works of the Convention (Malta Draft Articles, 1973), that the drafters sought to go beyond an anthropocentric understanding of the term 'environment' and intended it to include the entire marine ecosystem, especially the habitats of species in areas which often contain UCH (Czybulka, 2017, mn 25). Thus, the obligation under Part XII guides how to address threats to the marine environment namely, deep sea mining and may even provide guidance to how to implement the duties under Articles 303 and 149 (see Varmer, 2020).

These central ideas inspire the duty to protect UCH, as contained in the 2001 Convention on the Protection of UCH and its predecessors.

2.5 Overview of the Treatment of UCH Prior to the 2001 UCH UNESCO Convention

The question of the protection of UCH was not broadly discussed in international public law prior to adopting the LOSC. In fact, before its adoption, the silence on this matter was so loud that it produced what some have called a 'legal labyrinth' (See Altes, 1976). Nevertheless, a few instruments had mentioned the importance of cultural heritage in general before the 2001 UCH Convention, even if just briefly.

As has been mentioned, the first international legal instrument that recognised the 'outstanding universal value' of both natural and cultural heritage was undoubtedly the 1972 World Heritage Convention. With it, Member States recognised it was essential 'to adopt new provisions in the form of a convention establishing an effective system of collective protection of the cultural and natural heritage of outstanding universal value, organized permanently and per modern scientific methods' (WHC, 1972, Preamble). The Stockholm Declaration (Stockholm Declaration, 1973) was also important in this process, calling for sustainable development and creating a framework meant to guide policymakers into making decisions that both fomented development and took into account the protection of the surrounding environment (Varmer, 2020, p. 88).

Later, in 1992, the adoption of the European Convention on the Protection of the Archaeological Heritage (Revised) embodied a spirit of protection of the 'archaeological heritage as a source of the European collective memory', considering as elements of this 'archaeological heritage all remains and objects and any other traces of mankind from past epochs (...) which help to retrace the history of mankind and its relationship with the natural environment' (European Convention on the Protection of the Archaeological Heritage, 1992, article 1). This idea was also affirmed in the 1992 Rio Declaration on Environment and Development, which established in Agenda 21 a duty to protect the marine environment, to cooperate for that purpose and an integrated management approach as well as a precautionary approach.[11] Although as Varmer notes, much of the focus is on the conservation of marine living resources, 'the consideration of cultural heritage can be found throughout, including environmental impact assessments and integrated management' (Varmer, 2020, p.88).

2.6 The 2001 Convention and Its Relationship with the Law of the Sea

The UNESCO 2001 Convention arose as the Agreement to implement the call for more details under LOSC Article 303(4) and consistent with Article 311. Before this Convention, the LOSC addressed UCH as 'objects of an archaeological and historical nature', a vague and controversial expression. It was only with the 2001 Convention that the expression Underwater Cultural Heritage was adopted and spread worldwide, seeking to describe 'all traces of human existence having a cultural, historical or archaeological character which have been partially or underwater, periodically or continuously, for at least 100 years such as: (i) sites, structures, buildings, artifacts, and human remains, together with their archaeological and natural context; (ii) vessels, aircraft, other vehicles or any part thereof, their cargo or other contents, together with their archaeological and natural context; and (iii) objects of prehistoric character' (UNESCO Convention on UCH, 2001, Article 1.1(a)).

At the negotiation meetings, there was consensus regarding four principles: 1) to protect and preserve UCH; 2) the preferred first policy option of *in situ* preservation and adherence to the Annex Rules when a party decides not to preserve *in situ,* and recovery is in the public interest; 3) no 'commercial exploitation' of UCH; and 4) cooperation among States to protect UCH, particularly for training, education, and outreach. The primary purpose and focus were to address the threat from activities directed at UCH, such as looting and unwanted salvage. However, some provisions

[11] Chapter 17.1 highlights how the LOSC 'sets forth rights and obligations of States and provides the international basis upon which to pursue the protection and sustainable development of the marine and coastal environment and its resources'.

pertain to other human activities, such as deep sea mining, that could indirectly damage or destroy UCH. For instance, the preamble of the Convention highlights 'the need to respond appropriately to the possible negative impact on the underwater cultural heritage of legitimate activities that may incidentally affect it'. Article 2 (3), is a reference to the duty to protect in Article 149 of the LOSC, placing a general duty of protection for all UCH even if there is no direct interest for a State in doing so. Furthermore, Article 8 of the 2001 UCH Convention provides that 'States Parties may regulate and authorize activities directed at underwater cultural heritage within their contiguous zone'. This provides some more detail on the implementation of the LOSC duty to protect, suggesting that it does include economic activities that may adversely impact UCH, namely salvage and mining. It may be argued that the practice of nations as reflected in the ICJ 2022 ruling is more likely that of the details set out in UNESCO convention than a strict and narrow interpretation of the LOSC article.[12] The Annex to the 2001 Convention, concerning activities directed at UCH, also addresses the threats to natural resources and the environment around it as, over time they become inextricably connected. Thus, hinting at the idea that UCH is, in fact, a part of the marine environment. For instance, its drafters stated in Rule 10 (l) that the project design required for all activities concerning UCH, mentioned in Rule 9, must contain an environmental policy. According to Rule 29, the project design must also be 'adequate to ensure that the seabed and marine life are not unduly disturbed'. This concern is again expressed in Rule 14, which states that 'The preliminary work referred to in Rule 10 (a) shall include an assessment that evaluates the significance and vulnerability of the underwater cultural heritage and the surrounding natural environment to damage by the proposed project, and the potential to obtain data that would meet the project objectives'.

The 2001 Convention definition of UCH expressly provides that UCH is not limited to objects, and the archaeological context, but the natural context or it is undeniable that UCH and natural heritage are linked, both in physical and legal terms. Thus, the 2001 Convention provides guidance for implementing the duty to protect under the LOSC, feeding the notion that UCH is part of a broad consideration of the impacts to the marine environment, and even a narrow reading requires the consideration of UCH when UCH also serves as an artificial reef, thus reinforcing the protection of Article 192 of the LOSC.

2.7 Intangible Cultural Heritage

Although a less discussed side of cultural heritage, intangible cultural heritage has found its way into the debate surrounding deep sea mining activities. Intangible cultural heritage was first acknowledged in the 1992 Rio Declaration on Environment and Development, in Principle 22, where it was recognised that 'Indigenous people

[12] See Alleged Violations of Sovereign Rights and Maritime Spaces in the Caribbean Sea (Nicaragua v. Colombia), Judgment, I.C.J. Reports 2022, p. 266, [180–185].

and their communities and other local communities have a vital role in environmental management and development because of their knowledge and traditional practices. States should recognize and duly support their identity, culture, and interests and enable their effective participation in achieving sustainable development'. It was only much later, however, that the UNESCO Intangible Cultural Heritage Convention of 2003 defined this concept as 'the practices, representations, expressions, knowledge, skills—as well as the instruments, objects, artifacts and cultural spaces associated in addition to that—that communities, groups and, in some cases, individuals recognize as part of their cultural heritage' (UNESCO Intangible Cultural Heritage Convention 2003, Article 2 (1)). Moreover, this instrument also reinforced the existing relationship between intangible cultural heritage and tangible cultural and natural heritage (UNESCO Intangible Cultural Heritage Convention 2003, Preamble). In the context of UCH, intangible cultural heritage usually refers to the relationships many communities have built with the ocean.

Take, for instance, the seafaring peoples who traversed the Pacific thousands of years ago, the methods of which some call sacred, or the sacred bond some cultures hold with the sea deeming it even an ancestor itself or a resting place for their ancestors. It is plausible that an activity as hazardous as DSM could hurt such communities. For a long time, this conversation has not found a place in the ongoing work of the ISA. Nevertheless, since 2023, these communities have started to have their voice heard amid negotiations for the exploitation regulations, seeking to include a definition of intangible underwater cultural heritage and methods for its protection in the ISA regulations. An outstanding achievement in this regard was the establishment of the Intersessional Working Group on Underwater Cultural Heritage (see Chap. 4, this vol.).[13] Although the work of this group is still in its early stages it shows much promise, putting hope into the hearts of those who yearned for it.

2.8 Applying the Precautionary Approach to Deep Sea Mining and the Protection of UCH

It is generally agreed upon that applying the precautionary approach presupposes that we are before three cumulative requirements. Firstly, the object of potential application of this approach must be an activity or substance. Secondly, said object must pose a risk of serious or irreversible harm. Thirdly, there must be scientific uncertainty about the degree, likeliness, or type of damage that can be caused by such an activity or substance (Recuerda, 2008, p. 9).[14]

[13] https://enb.iisd.org/international-seabed-authority-isa-council-28-2-12jul2023. See also https://www.isa.org.jm/wp-content/uploads/2024/06/Submission-UCH-Specific-Indigenous-representatives-organizations.pdf.

[14] Scientific uncertainty may arise due to a lack of data, the dubious origin of that data or even from contradicting data.

This approach has been included in several international legal instruments, such as the 1992 Rio Declaration, which is yet another instrument famous for its Principle 15 on the precautionary principle and Agenda 21 of specifically, Chapter 17.22, which calls for a precautionary approach to the protection of the marine environment, the 1995 Fish Stocks Agreement,[15] and, more recently, the Agreement on the Protection of Biodiversity Beyond National Jurisdiction, still to enter into force.

International jurisprudence, mainly that produced by the International Tribunal for the Law of the Sea (ITLOS), has also been an avid defender of the application of a precautionary approach to the protection of the marine environment, including deep sea mining. At ITLOS alone there have been three court cases, namely, the Southern Bluefin Tuna, the MOX Plant, and the Land Reclamation cases, and two Advisory Opinions, specifically that on state Responsibility concerning activities in the Area and that concerning IUU Fishing.

Thus, a path emerges where the precautionary approach is included in legal instruments and case law. It could become a common standard for hazardous activities, such as deep sea mining. As Kristina Gjerde puts it, 'to ensure consistent application of principles agreed to by the world community (…), the principles, policies, and best practices that were adopted (…) will need to be explicitly recognized and incorporated into management action at all levels' (Gjerde, 2006, p. 305).

The call for a moratorium regarding deep sea mining is perhaps the major manifestation of the application of the precautionary approach to this activity. This moratorium would not be an outright ban, but an opportunity to create the conditions for better and more informed policy decisions, namely concerning UCH. Conducting surveys to ensure that no UCH finds itself in planned exploitation areas, proper Environmental Impact Assessments, and significant natural and cultural heritage aside as Marine Protected Areas are but a few measures that could be implemented for this purpose. The alternative could be irreparable environmental harm and the further destruction of elements of Humanity's common culture and history.

References

Sources

Ago, R. (1977). Sixth report on state responsibility – The internationally wrongful act of the state, source of international responsibility (continued) (A/CN.4/302 and 1, 2 & 3).

Altes, A. (1976). Submarine antiquities: A legal labyrinth. *Syracuse Journal of International Law and Commerce, 4*, 77–96.

[15] The precautionary approach may also be found in the 1992 United Nations Framework Convention on Climate Change, the 1995 WTO Agreement on the Application of Sanitary and Phytosanitary Measures, the 1998 Agreement on the International Dolphin Conservation Program, the 2000 Convention on the Conservation and Management of Highly Migratory Fish Stocks in the Western and Central Pacific Ocean, the 2001 International Convention on the Control of Harmful Anti-Fouling Systems, and the 2001 Agreement on the Conservation of Albatrosses, to name a few.

Aznar, Mariano & Varmer, Ole. (2013). The Titanic as Underwater Cultural Heritage: Challenges to its Legal International Protection. *Ocean Development and International Law*. 44 (1).

Aznar, M. J. (2014). The contiguous zone as archaeological zone. *The International Journal of Marine and Coastal Law, 29*, 1–51.

Aznar, M. J. (2022). Comentário ao artigo 303° objetos arqueológicos e históricos achados no mar. In W. Brito, *Comentários à Convenção das Nações Unidas sobre Direito do Mar* (pp. 567–593). Ed. Almedina.

Blanchard, C., et al. (2023). The current status of deep-sea mining governance at the International Seabed Authority. *Marine Policy, 147*.

Churchill, R.R. , et al. (2022). The Law of the Sea. Oxford University Press.

Czybulka, D. (2017). Art. 192. In A. Proelss (Ed.), *United Nations Convention on the Law of the Sea: A Commentary (A. R. Maggio, E. Blitza, & O. Daum, Assistant Editors)*. Nomos Verlagsgesellschaft.

Ermida, M. P. (2024). The duty to protect our ocean heritage from bottom trawling. In C. Jarvis (Ed.), *Threats to our ocean heritage: Bottom trawling*. SpringerBriefs in Archaeology, Springer.

Gjerde, K. (2006). High seas fisheries management under the Convention on the Law of the Sea. In D. Freestone, R. Barnes, & D. M. Ong (Eds.), *The law of the sea: Progress and prospects* (pp. 281–307). Cambridge University Press.

Heafy, E. (2014). Access and benefit-sharing of marine genetic resources from areas beyond national jurisdiction: Intellectual property—Friend, not foe. *Chicago Journal of International Law, 14*(2), 508.

International Seabed Authority. (2020). *Comments on the draft regulations on the exploitation of mineral resources in the area (ISBA/26/C/L.1)*. Retrieved December 30, 2024, from https://www.isa.org.jm/wp-content/uploads/2022/06/isba-26c-11-en_2.pdf.

Kiss, A. (1985). The common heritage of mankind: Utopia or reality? In *The library of essays in international law* (Vol. 40, pp. 423–441). Ashgate.

Oxman, B. (1988). Marine archaeology and the international law of the sea. *Columbia Journal of Law, 12*, 353–372.

Pardo, A. (1975). *The common heritage: Selected papers on oceans and world order 1967-1974*. Malta University Press.

Pardo, A. (1967). *Speech at United Nations General Assembly Twenty-Second Session Official Records (A/6695; A/C.1/952)*. United Nations General Assembly.

Pureza, J. M. (1998). *O património comum da humanidade: Rumo a um direito internacional da solidariedade*. Edições Afrontamento.

Recuerda, M. A. (2008). Dangerous interpretations of the precautionary principle and the foundational values of European Union food law: Risk versus risk. *Journal of Food and Law Policy, 4*(1), 1–42.

Risvas, M. (2013). The duty to cooperate and the protection of underwater cultural heritage. *Cambridge Journal of International and Comparative Law, 2*(3), 562–590.

Scovazzi, T. (2004). Mining, protection of the environment, scientific research, and bioprospecting: Some considerations on the role of the International Seabed Authority. *The International Journal of Marine and Coastal Law, 19*(4), 383–409.

Scovazzi, T. (2007). The concept of common heritage of mankind and the genetic resources of the seabed beyond the limits of national jurisdiction. *Agenda Internacional, 14*(25), 11–24.

Scovazzi, T. (2017a). Article 149. In A. Proelss (Ed.), *United Nations Convention on the Law of the Sea: A Commentary (A. R. Maggio, E. Blitza, & O. Daum, Assistant Editors)*. Nomos Verlagsgesellschaft.

Scovazzi, T. (2017b). Article 303. In A. Proelss (Ed.), *United Nations Convention on the Law of the Sea: A Commentary (A. R. Maggio, E. Blitza, & O. Daum, Assistant Editors)*. Nomos Verlagsgesellschaft.

Tanaka, Y. (2019). *The international law of the sea* (3rd ed.). Cambridge University Press.

Varmer, O. (2020). The duty to protect underwater cultural heritage and to cooperate for that purpose under law and policy. In H. Karan, X. Ademuni-Odeke, & K. Türk (Eds.), *The legal regime of underwater cultural heritage and marine scientific research* (pp. 77–117).
Vönecky, S., & Höfelmeier, C. (2017). Article 136. In A. Proelss (Ed.), *United Nations Convention on the Law of the Sea: A commentary*. Nomos Verlagsgesellschaft.
Wolfrum, R. (1983). The principle of the common heritage of mankind. *Heidelberg Journal of International Law, 43*.
Zhang, H. (2013). The sponsoring state's "obligation to ensure" in the development of the international seabed area. *The International Journal of Marine and Coastal Law, 28*(4).

Cases

Alleged Violations of Sovereign Rights and Maritime Spaces in the Caribbean Sea (Nicaragua v. Colombia), Judgment, I.C.J. Reports 2022, p. 266.
Land Reclamation in and around the Straits of Johor (Malaysia v. Singapore), Provisional Measures, Order of 8 October 2003, ITLOS Reports 2003, p. 10.
MOX Plant (Ireland v. United Kingdom), Provisional Measures, Order of 3 December 2001, ITLOS Reports 2001, p. 95.
Responsibilities and obligations of States with respect to activities in the Area, Advisory Opinion, 1 February 2011, ITLOS Reports 2011, p. 10.
Request for Advisory Opinion submitted by the Sub-Regional Fisheries Commission, Advisory Opinion, 2 April 2015, ITLOS Reports 2015, p. 4.
Southern Bluefin Tuna (New Zealand v. Japan; Australia v. Japan), Provisional Measures, Order of 27 August 1999, ITLOS Reports 1999, p. 280.

Conventions and Agreements

Agreement under the United Nations Convention on the Law of the Sea on the conservation and sustainable use of marine biological diversity of areas beyond national jurisdiction. 19 June 2023. Not yet in force. The Agreement shall enter into force 120 days after the date of deposit of the sixtieth instrument of ratification, approval, acceptance or accession. https://treaties.un.org/Pages/ViewDetails.aspx?src=TREATY&mtdsg_no=XXI-10&chapter=21&clang=_en.
Convention on the Protection of the Underwater Cultural Heritage. Adopted 2 November 2001. Entered into force 2 January 2009. https://www.unesco.org/en/underwater-heritage/2001-convention.
Convention for the Safeguarding of the Intangible Cultural Heritage. Adopted 17 October 2003. Entered into force 20 April 2006. https://ich.unesco.org/en/convention#art2.
1982 United Nations Convention on the Law of the Sea – Treaty Status. https://treaties.un.org/pages/ViewDetailsIII.aspx?src=TREATY&mtdsg_no=XXI-6&chapter=21&Temp=mtdsg3&clang=_en#:~:text=Signatories%20%3A%20157.,Parties%20%3A%20169.&text=CTC%2DArabic%3B%20CTC%2DChinese,1833%2C%20p.

Chapter 3
Protecting Underwater Cultural Heritage Under the BBNJ Agreement

Duncan Currie

Abstract Deep sea mining (DSM) has the potential to impact underwater cultural heritage, both tangible and intangible. The BBNJ Agreement will have implications for any underwater cultural heritage affected by DSM activities under its focus on marine biodiversity, its provisions on EIAs including strategic environmental assessments (SEAs) and ABMTs, and for its specific recognition of cultural values. It also has implications for the positions of BBNJ Parties when participating in the International Seabed Authority (ISA). Traditional knowledge is mainstreamed throughout the BBNJ Agreement, as is consultation and transparency. Protecting underwater cultural heritage under the BBNJ Agreement is likely to involve extensive consultation with both the ISA and UNESCO. Cultural matters as well as traditional knowledge are specifically included in both the ABMT and the EIA parts of the Agreement, but are also likely to arise in other contexts, including capacity building and the transfer of marine technology. Once BBNJ enters into force, international cooperation provisions including with respect to the ISA where activities in the Area are concerned and UNESCO when underwater cultural heritage or intangible cultural heritage is concerned will be important in resolving issues.

Keywords Deep sea mining · BBNJ treaty · Environmental law · International seabed authority · Cultural heritage

D. Currie (✉)
Globelaw, Christchurch, New Zealand
e-mail: duncanc@globelaw.com

C. Jarvis (ed.), *Threats to Our Ocean Heritage: Deep Sea Mining*,
SpringerBriefs in Archaeology, https://doi.org/10.1007/978-3-031-98238-5_3

3.1 Introduction and Context

Deep sea mining (DSM) has the potential to impact underwater cultural heritage, both tangible and intangible. The international community adopted the Agreement under the United Nations Convention on the Law of the Sea on the conservation and sustainable use of marine biological diversity of areas beyond national jurisdiction (BBNJ Agreement) in June 2023. The objective of the BBNJ Agreement is to ensure the conservation and sustainable use of marine biological diversity of areas beyond national jurisdiction (ABNJ) for the present and in the long term, through effective implementation of the relevant provisions of the Convention and further international cooperation and coordination. However, the BBNJ Agreement will also address cultural values in ABNJ, in the context of area-based management tools (ABMTs) including marine protected areas (MPAs) and environmental impact assessments (EIAs). It also includes the principle of the common heritage of humankind which is set out in the 1982 Law of the Sea Convention in Article 7(b).

The BBNJ Agreement will also have implications for any underwater cultural heritage affected by DSM activities under its focus on marine biodiversity, its provisions on EIAs including strategic environmental assessments (SEAs) and ABMTs, and for its specific recognition of cultural values. It also has implications for the positions of BBNJ Parties when participating in the International Seabed Authority (ISA) and other legal instruments and frameworks and relevant global, regional, subregional and sectoral bodies (IFBs).

The BBNJ Agreement preamble notes that Parties desire to act as stewards of the ocean in ABNJ on behalf of present and future generations by protecting, caring for and ensuring responsible use of the marine environment, maintaining the integrity of ocean ecosystems, and conserving the inherent value of biodiversity of areas beyond national jurisdiction. The BBNJ Agreement also embraces transparency in a stand-alone Article 48, which includes facilitation of the participation of, and consultation with, Indigenous Peoples and local communities with relevant traditional knowledge, the scientific community, civil society, and other relevant stakeholders. Throughout the Agreement, traditional knowledge as well as transparency is embraced and mainstreamed.[1] Given that deep sea mining may impact Indigenous People's culture, the BBNJ focus on traditional knowledge has significance for deep sea mining (Mulalap et al., 2020). Similarly, the BBNJ Agreement in Article 44, in the context of capacity building and the transfer of marine technology, lists information dissemination and awareness-raising, including with respect to relevant traditional knowledge of Indigenous Peoples and local communities, in line with the free, prior, and informed consent of these Indigenous Peoples and, as appropriate, local communities. Indigenous cultural relationships to the seabed (Perez-Alvaro, 2023) have been taken into account in the context of DSM decisions in South Africa (Sustaining the Wild Coast v Minister of Mineral Resources and Energy, 2022) and New Zealand (Trans-Tasman Resources Ltd v Taranaki-Whanganui Conservation

[1] Traditional knowledge is referred to no less than 29 times in the BBNJ Agreement.

Board, 2021).[2] The issue of intangible cultural heritage has arisen in numerous contexts in deliberations of the ISA (see Chap. 4, this vol.).

The BBNJ Agreement applies to ABNJ (Article 3): the high seas and the Area (Article 1(2)). The Area is defined in the Convention to mean the seabed and ocean floor and subsoil thereof beyond the limits of national jurisdiction. Activities in the Area shall be organised, carried out, and controlled by the Authority on behalf of mankind as a whole under Article 153(1) of the LOSC. As such, the BBNJ Agreement does not apply directly to the continental shelf, including the extended continental shelf, but does apply to the high seas above an extended continental shelf, which can extend up to 150 Nautical Miles beyond an exclusive economic zone (EEZ).

In 2020, IUCN resolution 122 called for a moratorium on deep sea mining, a call which has since been joined by 37 States calling for a moratorium or precautionary pause (IUCN, 2020). This chapter addresses the relationship between BBNJ and the ISA as it stands today.

The United Nations Educational, Scientific and Cultural Organization (UNESCO) in terms of BBNJ is a relevant IFB with respect to tangible and intangible cultural heritage. The UNESCO Director-General is the Secretariat of the UNESCO Convention on the Protection of the Underwater Cultural Heritage. The UNESCO Convention for the Safeguarding of the Intangible Cultural Heritage likewise is assisted by the UNESCO Secretariat.

3.2 Requirements of the BBNJ Agreement

The BBNJ Agreement has a strong emphasis on traditional knowledge. In Article 13 it requires that Parties shall take legislative, administrative or policy measures, with the aim of ensuring that traditional knowledge associated with marine genetic resources (MGRs) in areas beyond national jurisdiction that is held by Indigenous Peoples and local communities shall only be accessed with the free, prior and informed consent or approval and involvement of these Indigenous Peoples and local communities. Such MGRs may be found in the deep-sea Area impacted or potentially impacted by deep sea mining.

Access to such traditional knowledge may be facilitated by the Clearing-House Mechanism. Access to and use of such traditional knowledge shall be on mutually agreed terms under Article 13. Thus, Indigenous Peoples and local communities when faced with issues relating to traditional knowledge relating to MGRs would be able to turn to the Clearing-House Mechanism or indeed the Conference of the Parties or the implementation and compliance mechanism.

[2] See paras 9, 172, noting that considering the proposed activity in terms of tikanga may indicate that material harm extends beyond the physical effects of a discharge, or that pollution can be spiritual as well as physical.

The BBNJ Agreement in Article 7 sets out general principles and approaches by which Parties shall be guided in order to achieve the objectives of the Agreement. Notable in the heritage context are the principle of the common heritage of humankind which is set out in the Convention (Article 7(b)) and an integrated approach to ocean management as well as the freedom of marine scientific research, together with other freedoms of the high seas. Also notable are the use of relevant traditional knowledge of Indigenous Peoples and local communities, where available and the respect, promotion and consideration of their respective obligations, relating to the rights of Indigenous Peoples or of local communities when taking action to address the conservation and sustainable use of marine biological diversity of areas beyond national jurisdiction. The BBNJ also contains numerous provisions requiring consultation with Indigenous Peoples and local communities, described below.

3.3 Area-Based Tools and Marine Protected Areas

Area-based management tools and marine protected areas are provided for in Part III of the BBNJ Agreement. The BBNJ Conference of the Parties (CoP) is mandated to establish ABMTs including MPAs (Article 19). Consultations would need to be held with the ISA when a proposed ABMT involves the international seabed Area (Article 19(2)), as well as UNESCO when underwater cultural heritage or intangible cultural heritage are involved, and the views of these IFBs (Article 21(1)) invited including on the proposal's merits, scientific input, existing measures adopted by the ISA for the relevant area or for adjacent areas, views regarding any aspects of the measures and other elements for a draft management plan identified in the proposal that fall within the competence of the ISA, and views regarding any relevant additional measures that fall within the competence of the ISA.

Article 25(1) of the BBNJ Agreement provides that Parties must ensure that activities under their jurisdiction or control that take place in ABNJ are conducted consistently with the decisions adopted under Part III. This would, for example, apply to sponsoring States which would need to ensure activities taking place in areas under contract with the ISA are consistent with ABMTs, including MPA decisions made under Part III. So, where an MPA objective includes the protection of cultural values, Parties to BBNJ need to act both within the ISA and bodies such as UNESCO to this end (Article 17(d)).

More specifically, under Article 25(4), Parties are to promote, as appropriate, the adoption of measures within bodies, which would again include the ISA and UNESCO, to support the implementation of the decisions and recommendations made by the Conference of the Parties under Part III. This could mean that BBNJ Parties would need to promote the adoption of measures in the ISA and UNESCO to protect cultural values. Cultural factors are specifically included in Annex I as indicative criteria for the identification of areas. So, a proposal for an MPA could be made by a Party or Parties based at least in part on cultural values present in an Area.[3]

Parties are also to encourage non-Parties to adopt measures to support CoP decisions and recommendations on MPAs (Article 25(5)). Note that Indigenous People and local communities are to be consulted over any proposals for ABMTs (Article 19(2)), and that proposals shall be formulated on the basis of the best available science and scientific information and, where available, relevant traditional knowledge of Indigenous Peoples and local communities, taking into account the precautionary approach and an ecosystem approach (Article 19(3)). ABMT proposals are to include human activities in the area, including uses by Indigenous Peoples and local communities, and their possible impact, if any (Article 19(4)(c)), as well relevant scientific input and, where available, traditional knowledge of Indigenous Peoples and local communities (Article 19(4)(j)). Likewise, Indigenous People and local communities are to be consulted on submitted proposals (Article 21(1) and see Article 21(2)(c)).

The BBNJ CoP may take decisions on measures compatible with those adopted by the ISA, in cooperation and coordination with the ISA (Article 22(1)(b)). It may, where proposed measures are within the competence of the ISA, make recommendations to Parties to both BBNJ and to LOSC to promote the adoption of relevant measures through the ISA in accordance with its mandate (Article 22(1)(c)). In other words, the BBNJ CoP may recommend that BBNJ Parties, which are also parties to LOSC, promote the adoption of measures which will promote the objectives of the MPA, in accordance with the ISA's mandate. The same applies to parties to UNESCO with respect to heritage issues.

The BBNJ CoP is required to respect the competence of, and not undermine, bodies such as the ISA and UNESCO (Article 22(2)). The competence of the ISA is defined in terms of activities in the Area—meaning all activities of exploration for, and exploitation of, the resources of the Area.[4] The CoP is also to make arrangements for regular consultations to enhance cooperation and coordination with the ISA as well as coordination with regard to related measures adopted by the ISA and other IFBs such as UNESCO (Article 22(3)).

Where new MPAs are proposed and the proposal includes the Area, the ISA should be consulted (Article 19(2)) and invited by the BBNJ Secretariat to submit views, scientific input, information on existing measures and views on other matters (Article 21(2)(b)). As noted earlier, when underwater cultural heritage and/or intangible cultural heritage are at issue, UNESCO should likewise be consulted and invited to submit views. The CoP may take decisions on measures compatible with

[3] For proposals see BBNJ Agreement Art. 19.

[4] The competence of the ISA is provided in Article 153(1), which provides that 'activities in the Area shall be organized, carried out and controlled by the Authority on behalf of mankind as a whole in accordance with (Article 153) as well as other relevant provisions of (Part XI) and the relevant Annexes, and the rules, regulations and procedures of the Authority'. See 1982 Convention Art. 1(3). 'Resources' means all solid, liquid or gaseous mineral resources in situ in the Area at or beneath the seabed, including polymetallic nodules: Convention Art. 133(a).

those adopted by the ISA, in cooperation and coordination with the ISA (Article 22(1)(b)). The CoP may also, where proposed measures are within the competence of the ISA, make recommendations to BBNJ Parties and to the ISA to promote the adoption of relevant measures through the ISA, in accordance with its mandate (Article 22(1)(c)). The CoP shall respect the competence of, and not undermine the ISA, as well as UNESCO, as IFBs (Article 22(2)).

An issue may arise with respect to damage to underwater cultural heritage and/ or intangible cultural heritage caused or threatened by mining activities directly on the seabed or by sediment plumes generated by any deep sea mining and occurring in the water column. If such an activity or proposal was brought to the attention of the CoP, one response could be to take emergency measures, under Article 24, but only if it was or threatened to be as a human-caused disaster which has caused, or is likely to cause, serious or irreversible harm to marine biological diversity of ABNJ (Article 24(1)). Any such decision could only be taken if, following consultation with the ISA and UNESCO, the serious or irreversible harm cannot be managed in a timely manner through the application of the other Articles of the BBNJ Agreement or by the ISA or UNESCO (Article 24(2)).

3.4 Environmental Impact Assessments

The BBNJ Agreement operationalises the requirements of Article 206 of LOSC by establishing processes, thresholds and other requirements for conducting and reporting assessments by Parties to ensure that activities covered by Part IV are assessed and conducted to prevent, mitigate and manage significant adverse impacts for the purpose of protecting and preserving the marine environment (Article 27(b)). Article 31 requires that relevant traditional knowledge of Indigenous Peoples and local communities is taken into account during screening, scoping and impact assessment and evaluation, as well as that social and cultural impacts are to be included in EIA scoping reports, as well as reviews under Article 37, and Article 35 requires monitoring of social and cultural impacts.

Article 29 of the BBNJ Agreement specifically provides for the relationship between the BBNJ Agreement and EIA processes under bodies such as the ISA. The CoP is to establish mechanisms for cooperation between the Scientific and Technical Body (STB) and IFBs such as the ISA (Article 29(2)).

Secondly, in an intensively negotiated provision, the Agreement provides that if the Party with jurisdiction or control over a proposed activity (such as a sponsoring State, being a member State that allows a private entity to conduct seabed mining activities on its behalf) determines that the impacts have been assessed in accordance with the requirements of a body such as the ISA and that either:

(i) the assessment already undertaken for the planned activity is equivalent to the one required under Part IV, and the results of the assessment are taken into account; or

(ii) the regulations or standards of the relevant legal instruments or frameworks or relevant global, regional, subregional or sectoral bodies arising from the assessment were designed to prevent, mitigate or manage potential impacts below the threshold for environmental impact assessments under this Part, and they have been complied with, then a BBNJ EIA need not be completed. The first requires an equivalence test. As EIAs are by their nature procedural, it seems logical that at least a procedural equivalence test be applied.

The second test requires that there are regulations or standards in place, that they were designed to prevent, mitigate or manage potential impacts below the threshold for environmental impact assessments under Part IV, and they have been complied with.

Regardless of whether an assessment passes either of the Article 29 tests, the Party concerned shall ensure that the EIA report is published through the BBNJ Clearing-House Mechanism. Unless the planned activities that meet the first test are subject to monitoring and review, Parties shall monitor and review the activities and for EIAs that pass the equivalence test, monitor and review and ensure that the monitoring and review reports are published through the Clearing-House Mechanism (Article 29(5) and (6)).

3.4.1 Regional Environmental Management Plans (REMPs)

The ISA sees regional environmental management plans (REMPs) as an essential element of its strategy for the protection of the marine environment, as reflected in its Strategic Plan for the period 2019–2023 (ISBA/24/A/10). Specifically, they are intended to provide area-based and other management tools and are intended to provide goals, rules, and management tools for a specific region where any mining could occur (ISA Council, 2022a). To date the only existing REMP is the environmental management plan (EMP) for the Clarion Clipperton Zone (CCZ) (ISA Legal and Technical Commission, 2011), and REMPs are under development for the North Mid-Atlantic Ridge (ISA Council, 2022b) and the Northwest Pacific and Indian Oceans.[5] The draft REMP for the North Mid-Atlantic Ridge would require that the REMP pays due regard to any human remains, archaeological or cultural objects as set out in Article 149 and relevant Authority regulations (ISA Council, 2022b, part III). No similar provision is included in the only REMP existing to date, the CCZ EMP. Under the CCZ EMP, the ISA has established as ABMTs Areas of Particular Environmental Interest (APEIs), seabed areas closed to mining activities,

[5] See ISA at https://www.isa.org.jm/protection-of-the-marine-environment/regional-environmental-management-plans/. See https://www.isa.org.jm/protection-of-the-marine-environment/regional-environmental-management-plans/other-regions/ for details of REMP development in regions other than the CCZ.

under its Clarion-Clipperton Zone Environmental Management Plan (Blanchard & Gollner, 2022).

3.4.2 Strategic Environmental Impact Assessments

Any strategic environmental impact assessment (SEA) (Article 39) conducted under BBNJ should assess the potential effects of plans or programmes, as well as of alternatives, on the marine environment (Article 39(1)), and should collate and synthesise the best available information about the area or region, assess current and potential future impacts and identify data gaps and research priorities (Article 39(2)), as well as take into account any relevant SEAs already carried out (Article 39(3)). A SEA should take into account cultural and heritage values.

3.5 Capacity Building and the Transfer of Marine Technology

Capacity building and the transfer of marine technology are provided in Part VI of the BBNJ Agreement and are intended to be supportive of the other parts of the Agreement (Article 40). International cooperation is an important way that the Part VI objectives are to be realised (Article 41), including with Indigenous Peoples and local communities as holders of traditional knowledge (Article 41(2)). A capacity-building and transfer of marine technology committee reports to the BBNJ CoP (Article 46). Where concerns arise, or may arise, with respect to cultural heritage, in the context of capacity building and the transfer of marine technology, they could be brought up in the committee by way of bringing a matter to the attention of the CoP.

3.6 The Relationship Between BBNJ and the ISA

An important provision relevant to the relationship between BBNJ and the ISA as well as other IFBs such as UNESCO is Article 5(2) which provides that the BBNJ Agreement shall be interpreted and applied in a manner that does not undermine IFBs and that promotes coherence and coordination with IFBs. The BBNJ Agreement has a strong focus on international cooperation and consultation and includes some innovative and important international cooperation provisions. International cooperation is provided for in Article 8, which provides that Parties shall cooperate under the BBNJ Agreement for the conservation and sustainable use of marine biological diversity of areas beyond national jurisdiction, including through strengthening and enhancing cooperation with and promoting cooperation among bodies. These would include such bodies as the ISA and UNESCO in the achievement of the objectives of the BBNJ Agreement. More specifically, Article 8(2) provides that Parties shall endeavour to promote, as appropriate, the objectives of the BBNJ Agreement when participating in decision-making in IFBs, again such as the ISA and UNESCO. So, when a question arises in BBNJ once it enters force with respect to either underwater cultural heritage, intangible cultural heritage or both, UNESCO would likely be a relevant IFB, and the consultation provisions of BBNJ would be applicable to UNESCO, as well as potentially to the ISA. BBNJ Parties would therefore need to promote BBNJ objectives, including with respect to IPLCs, in those bodies.

3.7 LOSC

Under Article 149 of the 1982 Convention, all objects of an archaeological and historical nature found in the Area shall be preserved or disposed of for the benefit of mankind as a whole, particular regard being paid to the preferential rights of the State or country of origin, or the State of cultural origin, or the State of historical and archaeological origin. This provision acts as a reinforcement of the BBNJ provisions on cultural values and impacts,[6] and it is notable that the State of cultural origin, as well as the State of origin or of historical and archaeological origin, be given particular regard.

3.8 Exploitation Regulations

The ISA is currently negotiating exploitation regulations, parts of which address underwater cultural heritage. But it is too early to assess what those provisions will be: Will there be an underwater cultural heritage committee? Will the regulations, if and when adopted, require free and prior informed consent by Indigenous People and local communities? Will the regulations recognise cultural heritage as such or only physical manifestations of cultural heritage?

[6] A similar provision in Article 303 for objects found at sea requires protection of objects of an archaeological and historical nature. Article 303 is without prejudice to other international agreements and rules of international law regarding the protection of objects of an archaeological and historical nature, and thus the two UNESCO Conventions are of relevance.

3.9 Conclusion

Protecting underwater cultural heritage under the BBNJ Agreement is likely to involve extensive consultation with both the ISA and UNESCO. Cultural matters as well as traditional knowledge are specifically included in both the ABMT and the EIA parts of the Agreement, but are also likely to arise in other contexts, including capacity building and the transfer of marine technology. Once BBNJ enters into force, international cooperation provisions including those with respect to the ISA where activities in the Area are concerned and UNESCO when underwater cultural heritage or intangible cultural heritage is concerned will be important in resolving issues.

References

Sources

Blanchard, C., & Gollner, S. (2022). Area-based management tools to protect unique hydrothermal vents from harmful effects from deep-sea mining: A review of ongoing developments. *Frontiers in Political Science, 4*, 1033251. https://doi.org/10.3389/fpos.2022.1033251

International Seabed Authority (ISA) Council. (2022a). *Report of the Chair of the Legal and Technical Commission on the work of the Commission at its twenty-seventh session* (No. ISBA/27/C/37). https://www.isa.org.jm/wp-content/uploads/2022/12/2212509E.pdf.

ISA Council. (2022b). *Regional environmental management plan for the Area of the northern Mid-Atlantic Ridge with a focus on polymetallic sulphide deposits* (No. ISBA/27/C/38). https://www.isa.org.jm/wp-content/uploads/2022/12/2212833E.pdf.

ISA Legal and Technical Comission. (2011). *Environmental Management Plan for the Clarion Clipperton Zone* (No. ISBA/17/LTC/7). https://www.isa.org.jm/wp-content/uploads/2022/06/isba-17ltc-7_0.pdf.

International Union for Conservation of Nature. (2020). *Protection of deep-ocean ecosystems and biodiversity through a moratorium on seabed mining* (WCC 2020 No. Res 122). https://portals.iucn.org/library/sites/library/files/resrecfiles/WCC_2020_RES_122_EN.pdf.

Mulalap, C. Y., Frere, T., Huffer, E., Hviding, E., Paul, K., Smith, A., & Vierros, M. K. (2020). Traditional knowledge and the BBNJ instrument. *Marine Policy, 122*, 104103. https://doi.org/10.1016/j.marpol.2020.104103

Perez-Alvaro, E. (2023). Indigenous rights and underwater cultural heritage: (de)constructing international conventions. *Maritime Studies, 22*(3), 31. https://doi.org/10.1007/s40152-023-00320-6

Cases

Sustaining the Wild Coast v Minister of Mineral Resources and Energy, Case 3491/2021 (High Court of South Africa September 1, 2022). https://groundup.org.za/media/uploads/documents/judgment_on_sustaining_wild_coast_v_minister_of_mineral_resources__energy__others.pdf.

Trans-Tasman Resources Ltd v Taranaki-Whanganui Conservation Board, NZSC 127 (Supreme Court of New Zealand September 30, 2021). https://www.courtsofnz.govt.nz/assets/cases/2021/2021-NZSC-127.pdf.

Conventions and Agreements

Agreement under the United Nations Convention on the Law of the Sea on the conservation and sustainable use of marine biological diversity of areas beyond national jurisdiction. 19 June 2023. Not yet in force. The Agreement shall enter into force 120 days after the date of deposit of the sixtieth instrument of ratification, approval, acceptance or accession. At https://treaties.un.org/Pages/ViewDetails.aspx?src=TREATY&mtdsg_no=XXI-10&chapter=21&clang=_en.

Convention on the Protection of the Underwater Cultural Heritage. Adopted 2 November 2001. Entered into force 2 January 2009. At https://www.unesco.org/en/underwater-heritage/2001-convention.

Convention for the Safeguarding of the Intangible Cultural Heritage. Adopted 17 October 2003. Entered into force 20 April 2006. https://ich.unesco.org/en/convention#art2 As at January 1, 2024 the 2003 Convention had 182 Parties. https://www.unesco.org/en/legal-affairs/convention-safeguarding-intangible-cultural-heritage#item-2.

1982 United Nations Convention on the Law of the Sea – Treaty Status. https://treaties.un.org/pages/ViewDetailsIII.aspx?src=TREATY&mtdsg_no=XXI-6&chapter=21&Temp=mtdsg3&clang=_en#:~:text=Signatories%20%3A%20157.,Parties%20%3A%20169.&text=CTC%2DArabic%3B%20CTC%2DChinese,1833%2C%20p.

Chapter 4
The International Seabed Authority and the Protection of Our Ocean Heritage

Mariano J. Aznar

Abstract The International Seabed Authority (ISA), established pursuant to the 1982 United Nations Convention on the Law of the Sea (LOSC) and the 1994 Implementation Agreement, is mandated to regulate mineral-related activities in the seabed beyond national jurisdiction, ensuring that such activities benefit humanity as a whole. Central to this mandate is the development of the Mining Code, a comprehensive framework governing exploration and exploitation activities. While regulations on exploration were adopted between 2010 and 2013, the adoption of exploitation regulations remains pending, despite the triggering of a 2-year deadline by Nauru in 2021. Environmental protection constitutes a key element of the Mining Code, including provisions for marine protected areas and independent environmental reviews. However, ongoing negotiations reflect significant divergences among States and stakeholders regarding environmental safeguards, the protection of underwater cultural heritage, and equitable benefit-sharing. The increasing demand for critical minerals, driven by technological innovation and the energy transition, intensifies the need for balanced regulatory measures. The ISA faces the complex challenge of reconciling sustainable economic development with the preservation of fragile deep-sea ecosystems amid rising geopolitical tensions. In this context, technological advances, public awareness, and adherence to the precautionary principle will be critical for the responsible governance of seabed resources.

Keywords International Seabed Authority · Underwater Cultural Heritage · Deepsea mining · Regulation of Exploitation · Marine environment · Sustainable development

M. J. Aznar (✉)
Universitat Jaume I, Castellón de la Plana, Spain
e-mail: maznar@uji.es

C. Jarvis (ed.), *Threats to Our Ocean Heritage: Deep Sea Mining*,
SpringerBriefs in Archaeology, https://doi.org/10.1007/978-3-031-98238-5_4

4.1 Introduction

The International Seabed Authority (ISA) is an autonomous international organisation established under the 1982 United Nations Convention on the Law of the Sea (LOSC),[1] to regulate activities on the seabed, ocean floor, and subsoil beyond the limits of national jurisdiction—referred to as the 'Area'. Part XI of the LOSC governs these activities, complemented by the 1994 Agreement relating to the Implementation of Part XI of LOSC (1994 Agreement),[2] which permitted not only the entrance into force of LOSC in 1994 but also the establishment of ISA. The ISA is headquartered in Kingston, Jamaica, and plays a central role in overseeing the exploration and exploitation of natural resources in the Area, which are considered the common heritage of humankind (LOSC, Article 136). Its primary mandate is to ensure that activities in the Area are carried out for the benefit of all humanity, with special consideration given to the needs of developing states (Salamanca Aguado, 2019). The authority is empowered to organise and control mineral exploration and mining, including the issuance of licenses and the enforcement of environmental safeguards to mitigate the impact of such activities on the marine ecosystem. Additionally, the ISA is tasked with promoting marine scientific research in the Area and ensuring equitable sharing of benefits derived from seabed resources. To achieve these goals, the organisation operates through a structured governance framework comprising of the Assembly, the Council, and the Secretariat, along with subsidiary bodies such as the Legal and Technical Commission (Blanchard et al., 2023).

While the Assembly's functions include approving the ISA's budget, electing members of the Council and subsidiary bodies, and formulating broad policies for activities in the Area, the Council is the executive organ of the ISA and consists of 36 member states, elected by the Assembly based on equitable geographic representation and stakeholder interests (e.g., major seabed resource consumers, investors, and coastal states). The Council is responsible for overseeing seabed activities, approving contracts for exploration and exploitation,[3] and developing detailed regulations known as the Mining Code, which are to be adopted by the Assembly. It

---

[1] Adopted 10 December 1982, entered into force 16 November 1994, 1833 UNTS 3. LOSC currently has 170 States parties plus the European Union, and relevant parts of its clauses have a customary nature, thus obliging even non-States parties (like Colombia, Iran or the US).

[2] Adopted 28 July 1994, entered into force 28 July 1996, 1836 UNTS 3.

[3] So far, ISA has entered into 15-year contracts for the exploration for polymetallic nodules, polymetallic sulphides, and cobalt-rich ferromanganese crusts with 22 contractors. Nineteen of these contracts are for the exploration for polymetallic nodules in the Clarion-Clipperton Fracture Zone (17) and Central Indian Ocean Basin (1), and Western Pacific Ocean (1). In addition, there are seven (7) contracts for exploration for polymetallic sulphides in the South West Indian Ridge, Central Indian Ridge and the Mid-Atlantic Ridge and five contracts for exploration for cobalt-rich crusts in the Western Pacific Ocean. For polymetallic nodules, the entitled exploration area allocated to each contractor is 75,000 km^2. For polymetallic sulphides, the entitled exploration area allocated to each contractor is 10,000 km^2 and consists of 100 blocks. Each block is no greater than 100 km^2. For cobalt-rich ferromanganese crusts, the entitled exploration area allocated to each contractor is 3000 km^2 and consists of 150 blocks. Each block is no greater than 20 km^2. More than 1.5 million km^2 of international seabed, roughly the size of Mongolia, has been set aside for mineral exploration. Source: ISA.

ensures that activities are carried out in line with LOSC and ISA's principles. Headed by the Secretary-General, the Secretariat is the administrative arm of the ISA. It manages the day-to-day operations, supports the other organs, and facilitates the implementation of the organisation's objectives. It also conducts research, provides technical assistance to member States, and oversees compliance with environmental and financial requirements. Finally, among the ISA subsidiary bodies, the Legal and Technical Commission (LTC) is an expert body, composed of individuals nominated by member States, which evaluates applications for exploration and exploitation licenses, reviews environmental impact assessments, and develops technical recommendations for the Council.

As the steward of vast and largely unexplored deep-sea regions, the ISA wields significant influence in shaping global maritime policies, exploring economic development opportunities while protecting the marine environment, and fostering international cooperation to address issues such as technological innovation, biodiversity protection, and equitable access to seabed resources. It is also responsible for developing a 'Mining Code', the regulations and standards to govern any potential deep-sea mining.

4.2 ISA Mining Code

The Mining Code is a comprehensive set of rules, regulations, and procedures developed by the ISA to govern the exploration and exploitation of mineral resources in the Area. The Mining Code is designed to ensure that activities in the Area are conducted in a manner that benefits all of humanity, protects the marine environment, and adheres to the principle of sustainable development, with a focus on assisting developing nations. It provides the legal and operational framework for how exploration and mining activities would be carried out, including guidelines on licensing, environmental protection, financial terms, and monitoring compliance.

The Mining Code is divided into regulations for different stages and types of activities. It includes provisions for the exploration of polymetallic nodules, polymetallic sulphides, and cobalt-rich ferromanganese crusts, which are rich in valuable metals like nickel, copper, cobalt, and rare earth elements. These resources have attracted significant interest due to their potential to support industries like renewable energy and advanced technologies. The Code also includes rules for contractors applying for exploration and exploitation licenses, detailing their rights and responsibilities, including the requirement to provide environmental impact assessments and comply with environmental management plans.

Between 2010 and 2013, the ISA adopted four different regulations on prospecting and exploration.[4] However, the adoption of regulations on exploitation —a

[4] See all the information and text of these regulations at ISA website, available at <https://www.isa.org.jm>.

much more sensitive issue—was forced by Nauru, sponsoring a Canadian mining company eager to begin exploitation in the Area: The Metals Company, Inc., via a wholly owned subsidiary Nauru Ocean Resources, Inc. (NORI). On 25 June 2021, Nauru made a formal request to ISA Council to complete the adoption of rules, regulations, and procedures necessary to facilitate the approval of plans of work for exploitation.[5] Triggering Article 15(b) of the 1994 Agreement Annex, the ISA Council had two years to complete the adoption of such regulations. Letter (c) of same clause establishes that if the Council did not complete the elaboration of the regulations within these two years, it should have to provisionally approve a plan of work for exploitation based on the existing provisions of LOSC and the 1994 Agreement, as well as the principle of non-discrimination among contractors.

The ISA deadline for drafting new regulation was set in July 2023, but the regulations on exploitation were not adopted due to the complexities of the issue and sound differences among negotiating States and potential contractors' interests.[6] Many countries have said they will not approve a plan of work to mine without finishing regulations on exploitation and the ISA Council, the body responsible for the decision, has taken two resolutions by consensus stating that no plans of work should be approved in the absence of regulations.

Actually, as this chapter is being written, draft regulations on exploitation are currently under negotiation at ISA Council. The LTC submitted a draft text in 2019, which has received numerous inputs and discussed by ISA States members and other stakeholders (international organisations, NGO, contractors). Informal working groups and intersessional working groups—including one on underwater cultural heritage—have been established by the Council to advance the work on the draft exploitation regulations, with a view to their adoption by 2025. A consolidated text was released on February 2024, containing a compilation of work carried out in the informal working groups of the Council (ISBA/29/C/CRP.1). However, there are still too many points under discussion (Pickens et al., 2024). If the natural marine environment *acquis* is more or less accepted and included in the regulations, no agreement has yet been reached on the mechanism of protection of underwater cultural heritage.

---

[5] "Letter dated 30 June 2021 from the President of the Council of the international Seabed Authority addressed to the members of the Council" (Doc. ISBA/26/C/38, 1 July 2021), available at <https://www.isa.org.jm/wp-content/uploads/2022/06/ISBA_26_C_38-2108753E.pdf>.

[6] Beyond the significant but still unknown impact on the marine environment, the economies of many nations and the livelihoods of indigenous groups who depend on the oceans for survival are to be assessed.

4.3 The Protection of Marine Natural Heritage

A key element of the Mining Code is its emphasis on environmental protection (Sun, 2023). It requires contractors to minimise the environmental impacts of their activities and establishes mechanisms for monitoring and mitigating potential damage to deep sea ecosystems, which are particularly fragile and not yet fully understood. This is why some States and most NGO involved in the negotiations have asked for a moratorium or a precautionary pause for the exploitation phase (Singh et al., 2025).[7] This trend against mining in the Area was revealed and increased during the adoption process of the BBNJ,[8] which incorporates various measures for the protection of the marine environment, some of which could be exported to mining regime in the Area (Robb et al., 2023; Friedman, 2024).[9]

Protection of the marine environment, including through the Mining Code, is a critical component of the ISA's mandate. To mitigate potential damage from mining operations, ISA regulations and guidelines include a comprehensive environmental approach as part of the Mining Code. These regulations are designed to strike a balance between enabling the sustainable use of seabed resources and safeguarding the marine environment for future generations. The key components of such approach contain several interlinked components, both theoretical and practical:

- Precautionary Approach: ISA adheres to the precautionary principle, particularly in light of the limited knowledge about deep sea ecosystems.[10]
- Monitoring and Reporting: Contractors must conduct continuous environmental monitoring during exploration and mining operations (Rojas & Philips, 2019). Data on water quality, sediment plumes, biodiversity, and ecosystem health must be reported to ISA. This information helps the Authority to assess compliance with environmental regulations and identify trends or emerging risks.
- Environmental thresholds and standards for mining activities, such as limits on sediment plume dispersal and noise pollution. These thresholds aim to minimise cumulative impacts on marine life and ecosystems. The ISA also encourages adaptive management, allowing regulations to evolve based on new scientific insights.
- Capacity Building for Developing States by the ISA, providing technical and financial support to help developing States enhance their capacity for conducting

[7] See more information at <https://deep sea-conservation.org/solutions/no-deep sea-mining/>.

[8] UN Convention on the Law of the Sea on the conservation and sustainable use of marine biological diversity of areas beyond national jurisdiction, adopted 4 March 2023, not yet in force. Information available at <https://treaties.un.org/doc/Publication/MTDSG/Volume%20II/Chapter%20XXI/XXI-10.en.pdf>.

[9] See Chap. 3 of this book.

[10] See ISA, No. 5 Discussion Paper: The Implementation of the Precautionary Approach by the International Seabed Authority (2017), available at <https://www.isa.org.jm/wp-content/uploads/2022/12/DP5.pdf>.

environmental assessments and monitoring. This ensures equitable participation in seabed governance and helps prevent environmental exploitation.

- Responsibility for Harm: Contractors are held accountable for any environmental harm caused by their activities. ISA regulations require financial guarantees or insurance to cover potential environmental damage and remediation costs. Responsibility also refers to States sponsoring contractors' activities.[11]
- Environmental Impact Assessments (EIA), identifying potential environmental (cumulative) impacts, including risks to biodiversity, ecosystem functioning, and water quality. EIA, once submitted to the ISA, must be reviewed by the LTC. They serve as a baseline for monitoring environmental changes during and after mining operations.
- Environmental Management Plans (EMP), prepared and implemented by contractors, outlining strategies for minimizing and mitigating environmental harm. EMP include measures to prevent pollution, protect biodiversity, and restore impacted areas where possible. ISA requires regular updates to EMPs based on ongoing monitoring and scientific findings.
- Preservation Reference Zones (PRZ), established to safeguard marine biodiversity. Contractors are required to set aside areas within their licensed regions as PRZ. These zones remain undisturbed and serve as control sites to measure the impact of mining on the surrounding environment.
- Marine Protected Areas (MPA) to be established in ecologically sensitive regions of the seabed. These areas would be completely off-limits to mining activities, providing refuge for species and habitats at risk.
- Independent Environmental Review, seeking the involvement of independent experts in reviewing EIA, EMP, and environmental monitoring data. This ensures transparency and accountability in the regulatory process.

4.4 The Protection of the Underwater Cultural Heritage

Underwater cultural heritage (UCH) located in the Area may benefit from advanced theoretical and practical instruments already present in the protection of marine natural heritage (Aznar, 2017). However, adopted Regulations on prospecting and exploring do not help. Whilst common Article 8 of these Regulations—applicable to prospectors—refers to any 'object of actual or potential archaeological or historical nature', regulations 35 (Nodules) and 37 (Sulphides and Crusts)—applicable to contractors—refer to 'any human remains of an archaeological or historical nature, or any object or site of a similar nature'. They do not follow either the terminology

[11] A sound discussion on responsibilities of States sponsoring contractors in the Area may be found in the advisory opinion gave by the International Tribunal for the Law of the Sea in 2011: *Responsibilities and obligations of States with respect to activities in the Area, Advisory Opinion, 1 February 2011, ITLOS Reports 2011*, 10 (available at <https://www.itlos.org/fileadmin/itlos/documents/cases/case_no_17/17_adv_op_010211_en.pdf>).

of LOSC ('archaeological and historical objects') or that of the UNESCO Convention on the Protection of the Underwater Cultural Heritage (2001 Convention),[12] which uses this later term of art (UCH), including 'human remains' therein.

As mentioned, an intersessional working group (WG) was created in 2023 to discuss the possible text of a regulation on exploitation addressing the protection of UCH in the Area (numbered DR35). Undeniably, it had to be framed within existing applicable law, both general (LOSC, Article 149) and particular (2001 Convention, Articles 11 and 12),[13] and taking into account Indigenous peoples' concerns regarding their cultural connection to the ocean.[14] Prior to the July 2024 meeting of the ISA Council, the WG discussed two sets of textual proposals: one on intangible UCH from representatives of Indigenous peoples and local communities from the Pacific, and one on tangible UCH contained in a proposal by Spain.[15] Despite efforts to find a common text, divergences on how to offer a workable proposal including both kind of heritage drive some like-minded States to move forward and suggest a draft text under three basic assumptions: *first*, that the future regulation cannot rewrite LOSC; *second*, that it should mainly focus on tangible UCH (including human remains and tangible objects associated to intangible cultural heritage); and, *third*, aligns as much as possible with the 2001 Convention. This last point, despite the still limited acceptance of the 2001 UNESCO Convention, received however a general support both in the WG and the July 2024 ISA Council meeting. Some delegates voiced their agreement with the general approach provided by Spain in their non-paper submitted in July 2024,[16] which proposes a system of notification between stakeholders (contractor, ISA Secretary-General, States members and competent organisations, most particularly UNESCO), a system of consultations among interested States (i.e. those with a verifiable link to UCH), and a decision-making process with their participation in the Council, taking into account the views of competent organisations. These decisions should follow archaeological applicable standards and adopted ISA guidelines,[17] and shall not affect existing rights of flag States upon their sunken State vessels and aircraft located in the Area. It is foreseen that contractors should not be entitled to compensation for any measure required in

[12] Adopted 2 November 2001, entered into force 2 January 2009, 2562 UNTS 1.

[13] While LOSC has been almost universally accepted, UCHC has been ratified only by 79 States by April 2025.

[14] See some approaches to the question of intangible cultural heritage in this chapter and Chap. 10 of this book.

[15] See Briefing Note for Thematic Discussion on "Underwater Cultural Heritage", available at <https://www.isa.org.jm/wp-content/uploads/2024/07/Briefing-Note-for-Thematic-Discussion-on-Underwater-Cultural-Heritage.pdf>.

[16] Spain's proposal was also supported by the main international scientific institution advising on UCH: the International Committee on Underwater Cultural Heritage of the International Council on Monuments and Sites (ICUCH-ICOMOS).

[17] It is under discussion, as proposed by Spain, the possibility of drafting specific guidelines for activities incidentally affecting UCH in the Area.

the regulation,[18] and that all information gathered should be registered for scientific purposes.

Delegates and NGOs also discussed the importance of intangible UCH,[19] with some pointing to its inextricable links to tangible UCH, while others said that the draft regulation should focus solely on tangible objects to avoid ambiguity.[20] It was generally agreed to explore the mechanism to attend the voices of indigenous peoples and local communities within the future regulations.[21]

4.5 Final Remarks

The main challenges before the ISA to adequately face the protection of our Ocean Heritage may be summarised around environmental concerns, including the fragility of deep sea ecosystems and possible biodiversity loss; the problems around how to incorporate intangible cultural heritage and cultural connection to the ocean; the still scientific uncertainty, which makes necessary a precautionary approach to mining in the Area; the complex balancing of economic interests with sustainability, whereas global resource demand for critical minerals increases to support renewable energy technologies and AI hardware; the necessary linkage of economic inequities and revenue-sharing mechanisms; or financial and economic benefits of seabed mining, particularly to assist landlocked and developing nations.

All these questions need to be peacefully nested in the Mining Code, as well as those related to monitoring, enforcement and compliance, and dispute resolution. Ongoing negotiation process develops in a global scenario where oceans are again conflicting areas. Competing interests and geopolitical rivalries fight in oceans as new disputed arenas.

---

[18] It is endorsed the idea that contractors should have taken measures to identify the presence of UCH during the exploration phase or during the preparation of the application for a Plan of Work; and, consequently, it has no right to compensation.

[19] See Chaps. 4 and 8 of this volume.

[20] Among the NGO, The Ocean Foundation (TOF) was particularly active suggesting ideas on how to address the protection of intangible UCH before future deep sea mining in areas of special relevance for indigenous peoples and local communities. See general TOF involvement at <https://oceanfdn.org/our-goals-for-the-international-seabed-authority-and-beyond/>.

[21] Indigenous peoples were mainly guided by the UN Declaration on the Rights of Indigenous Peoples (A/RES/61/295, 13 September 2007, Annex), most particularly its desire "to act as stewards of the ocean in areas beyond national jurisdiction on behalf of present and future generations", emphasising their rights "to free, prior, and informed consent before any deep-sea mining project which may affect intangible cultural heritage" (Statement circulated in the WG on UCH, on file with the author).

Last but not least, prolonged delays in the adoption of regulations may hinder investment and create uncertainty for contractors,[22] while rushing decisions risks compromising environmental and cultural heritage safeguards.

The ISA is at the centre of a delicate balancing act, seeking to promote the sustainable use of seabed resources while protecting the marine environment and ensuring equitable benefit-sharing. Addressing these challenges requires a comprehensive, science-based, and inclusive approach, along with stronger international cooperation and transparency—including public participation (Morgera & Lily, 2022)—for protecting our Ocean Heritage.

References

Aznar, M. J. (2017). Exporting environmental standards to the protection of underwater cultural heritage in the area. In J. Crawford, A. G. Koroma, S. Mahmoudi, & A. Pellet (Eds.), *The international legal order: Current needs and possible responses – Essays in Honour Djamchid Momtaz* (pp. 255–273). Brill/Nijhoff. https://doi.org/10.1163/9789004314375_019

Blanchard, C., Harrould-Kolieb, E., Jones, E., & Taylor, M. L. (2023). The current status of deep-sea mining governance at the International Seabed Authority. *Marine Policy, 147*, 105396. https://doi.org/10.1016/j.marpol.2022.105396

Dobush, B.-J., & Warner, M. (2024). *Deep sea mining isn't worth the risk: High costs, financial developments since 2021, and externalities stand to diminish theoretical returns on investments*. The Ocean Foundation. https://oceanfdn.org/wp-content/uploads/2024/02/dsm-finance-brief-2024.pdf.

Friedman, S. (2024). The interaction of the BBNJ agreement and the legal regime of the area, and its influence on the implementation of the BBNJ agreement. *Marine Policy, 167*, 106235. https://doi.org/10.1016/j.marpol.2024.106235

Morgera, E., & Lily, H. (2022). Public participation at the International Seabed Authority: An international human rights law analysis. *RECIEL, 31*(3), 374–388. https://doi.org/10.1111/reel.12472

Pickens, C., Lily, H., Harrould-Kolieb, E., Blanchard, C., & Chakraborty, A. (2024). From what-if to what-now: Status of the deep-sea mining regulations and underlying drivers for outstanding issues. *Marine Policy, 169*, 105967. https://doi.org/10.1016/j.marpol.2023.105967

Robb, S., Jaeckel, A., & Blanchard, C. (2023, April 13). How could the BBNJ Agreement affect the International Seabed Authority's Mining Code? *EJIL Talk!*. https://www.ejiltalk.org/how-could-the-bbnj-agreement-affect-the-international-seabed-authoritys-mining-code/.

Rojas, A., & Phillips, F.-K. (2019, February 27). *Effective control and deep seabed mining: Toward a definition*. Centre for International Governance Innovation. Liability Issues for Deep Seabed Mining Series. https://www.cigionline.org/publications/effective-control-and-deep-seabed-mining-toward-definition-1/.

Salamanca Aguado, M. A. (2019). La política ambiental de la Autoridad Internacional de los Fondos Marinos: el equilibrio entre la explotación y la conservación del patrimonio común de la humanidad. *Revista Catalana de Dret Ambiental, 10*(2), 1–43. https://doi.org/10.17345/rcda2684

[22] On 14 January 2025, a group of contractors send a letter to the ISA Council President expressing their 'concerns regarding the delay in the transition to the exploitation phase of activities in the Area and to urge the Council to complete the adoption of Exploitation Regulations by 2025' (file with the author).

Singh, P. A., Jaeckel, A., & Ardron, J. A. (2025). A pause or moratorium for deep seabed mining in the area? The legal basis, potential pathways, and possible policy implications. *Ocean Development & International Law, 56*(1), 18–44. https://doi.org/10.1080/00908320.2024.2439877

Sun, L. (2023). International seabed authority and its environmental mandate. In *International environmental obligations and liabilities in deep seabed mining* (pp. 57–109). CUP. https://doi.org/10.1017/9781108770125.005

Chapter 5
Archaeology in the Deep

Charlotte Jarvis

Abstract Shipwrecks found in the deep are valuable parts of Ocean Heritage. Thanks to technology, the deep sea (more than 200 m deep) has been reachable by archaeologists for decades. With future threats to Ocean Heritage such as deep sea mining, maritime archaeologists must leverage deep ocean research and visual media to engage the public and dispel myths about the deep sea. There must be interdisciplinary collaboration, particularly in light of the UN Decade of Ocean Science (2021–2030). Proper scientific baseline surveys, including archaeological analysis, are crucial for DSM projects to ensure the protection of Ocean Heritage. Establishing exclusion zones based on site-specific conditions and continuous monitoring are vital measures to safeguard cultural resources from dredging activities.

Keywords Underwater cultural heritage · Deep sea mining · Deep seabed mining · Deep-sea mining · Intangible cultural heritage · International seabed authority · Underwater archaeology

5.1 Introduction

Noted in the *Encyclopedia of Underwater and Maritime Archaeology* as 'the world's last great archaeological frontier' (Delgado, 1998, p. 127), shipwrecks found in the deep are valuable parts of Ocean Heritage. Thanks to technology, the deep sea (more than 200 m deep) has been reachable by archaeologists for decades. Tools such as Remotely Operated Vehicles (ROVs) and Autonomous Underwater Vehicles (AUVs) have aided in the search, survey, identification, and even excavation of underwater cultural heritage (UCH) sites (Søreide, 2011).

C. Jarvis (✉)
The Ocean Foundation, Washington, DC, USA

C. Jarvis (ed.), *Threats to Our Ocean Heritage: Deep Sea Mining*, SpringerBriefs in Archaeology, https://doi.org/10.1007/978-3-031-98238-5_5

The discovery of *Titanic* in 1985 is one of the first examples of a deep-water shipwreck survey. A joint team from Woods Hole Oceanographic Institution and the French Research Institute for Exploration of the Sea (IFREMER) used some of the most cutting-edge underwater vehicles of the time to search for the vessel. Robert Ballard and Jean Louis Michel headed up the teams and used the deep-towed new sonar system SAR (System Acoustique Remorquè) and an unmanned deep-sea vehicle (*Argo*) to search the seafloor. They found a debris trail which led them to the wreck in 3800 m of water 300–360 nautical miles off the coast of Newfoundland, Canada. In 1987, another submersible, *Nautile*, equipped with a robotic arm controversially recovered the first artefacts from the seafloor.

Anna Marguerite McCann and Robert Ballard, as part of The JASON Project, were also among the first major users of remotely operated technology for maritime archaeology in 1989 for the Skerki Bank project, which investigated a late Roman wreck off the coast of Sicily in more than 700 m of water (McCann & Oleson, 2004). Ballard founded the Institute for Exploration at Mystic Aquarium in Connecticut with temporary facilities at the University of Rhode Island and through these institutions conducted further expeditions in Skerki Bank and in the Black Sea using ROVs into the early 2000s (Brennan et al., 2018).

Remote work on deep sites became more accessible in 2008 when two ships of exploration were launched (Brennan et al., 2018). The US National Oceanic and Atmospheric Administration's (NOAA's) ship *Okeanos Explorer* and Ballard's Ocean Exploration Trust's E/V *Nautilus* were outfitted with equipment and permanent telepresence technologies to conduct multiple expeditions each year with the goal of year-round live streams of the work being done onboard (Brennan et al., 2018). The ships began work from the Mediterranean to the Pacific with live feeds sent to shore (see Brennan et al., 2018 for a detailed discussion of the vessels' missions). *Okeanos Explorer* and *Nautilus* became tools 'in the arsenal of deep-water archaeologists that allows access to sites out of reach' (Brennan et al., 2018, p. 103). Their telepresence capability allows archaeologists and other marine scientists to work together to uncover the stories of the deep, but it is not currently required by International Seabed Authority (ISA) regulations.[1]

Deep-water archaeology continues to evolve and improve, and several excavations and surveys have taken place over the last two decades. The deep sea is not as inaccessible to archaeologists as it used to be. *Titanic* was mapped in 2010 by an ROV with high-resolution sonar and two highly sophisticated camera systems allowing the shipwreck to finally 'be treated and explored like any other underwater site', showing that 'even extreme depth is no longer an obstacle to archaeology' (Delgado, 2012).

[1] ISA Technical Study no. 29 on 'Remote monitoring systems in support of inspection and compliance in the Area' sees this as part of the 'Optimal' and 'Maximum' option. This would require the contractor to conduct 'visual monitoring on and near the seabed during mining' with 'Observations from ROV/AUV camera on biological communities, substrate and bottom topography (video transects)' (p. 50). The recommendations of this study had not been discussed by the Council or reflected in the Regulations and Procedures (RRPs) as of November 2024 (Pickens et al., 2024).

But why are these sunken vessels important not only to maritime archaeologists and historians but also to the health of the ocean? Why do these crucial sites of Ocean Heritage need to be protected from deep sea mining? For archaeologists, deep-water wrecks often have remarkable preservation due to the lack of oxygen, the lack of disturbance by anthropogenic activities, the lack of physical damage from oceanic processes such as waves and storm surge, and their relative inaccessibility which deters most treasure hunters and salvors from pillaging the wrecks (Beltrame et al., 2022). This final reason is crucial for systematic, scientific study of wrecks. This chapter does not focus on finding cultural heritage that would be in direct risk from deep sea mining (DSM) operations in specific locations. Instead, the following case studies show that deep-water archaeology has the potential to, and track record of, revealing well-preserved vessels that fill gaps in our historical knowledge. It should be considered during mining regulation discussions and considerations.

The logistics of any form of maritime archaeology creates challenges, and science in the deep-water is inherently more complex. The marine environment is a difficult place to work. Weather can change quickly, salt water corrodes instruments as soon as they are submerged, and the deeper the work is, the more pressure is exerted on operations and machinery. Scientific investigation in the deep must rely on remote sensing technologies to gather data. This increased reliance on technology and instruments for data collection provides more opportunities for potential equipment failure. Archaeology has not traditionally been a field that is known for its cutting-edge technology and projects often lack sufficient funding to embrace the newest tools. However, in the remote sensing field, 'marine archaeology was an early adaptor of such technology, and has used side-scan sonar (SSS), multibeam echosounder (MBES), and sub-bottom profilers (SBP) to detect, map, and monitor cultural heritage on the seabed for decades' as well as using marine magnetometers (Ødegård et al., 2018, p. 2). While there are still challenges, the following case studies show that UCH in the deep can also be scientifically studied and the technology already exists to do so (Fig. 5.1).

Another key importance of these deep-water sites is their role as ecosystems. Shipwrecks often function as artificial reefs and, in many places, they are the only hard substrate on an otherwise barren seafloor (Pearson & Thompson, 2023; Krumholz & Brennan, 2015; Meyer-Kaiser et al., 2022; Hampel et al., 2022). The hard substrate acts as an anchor for life to colonise and grow. Such wrecks also act as shelters for juvenile fish species and are as crucial to the seafloor ecology as natural coral reefs or seamounts (Paxton et al., 2024). These shipwrecks cannot be viewed only as examples of UCH: they are also part of the natural ocean landscape and our Ocean Heritage.

The way we speak about the deep ocean as scientists influences public perception. The deep is often characterised as an unknown, dark, foreign place which creates a (sub)conscious psychological response. It conjures up the image of a bleak, barren land, an 'other' place which leads to an out of sight, out of mind response. While this language has become common place in how we discuss the sea in pop culture media (seen in shows such as *Mysteries of the Deep*), it is in no way an accurate description of the deep, or of the work that scientists are currently

Fig. 5.1 An ROV carefully excavated on the Mardi Gras Wreck in the Gulf of Mexico at 1219 m of water. (Source: BOEM)

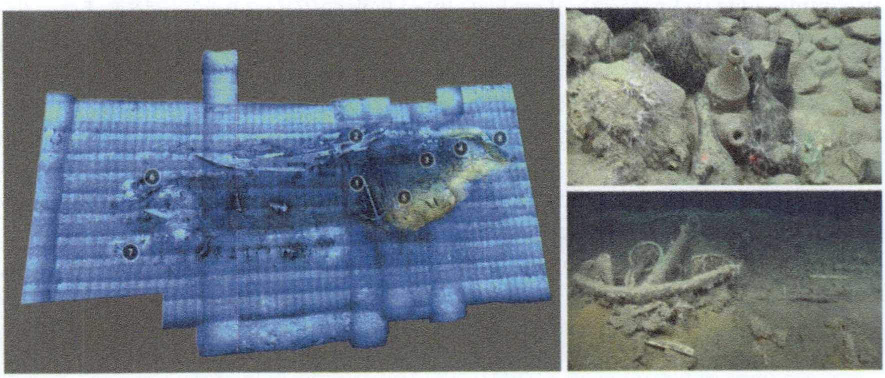

Fig. 5.2 BOEM's virtual archaeology museum is a great way to bring the mysteries of underwater archaeology to the public. The website shows 3D models of shipwreck sites, with annotation, and further images of features on the wrecks. (Source: BOEM, https://www.boem.gov/environment/virtual-archaeology-museum)

conducting far below the surface. Out of sight should not necessarily equate to out of mind (Fig. 5.2). Although high resolution data and locations of archaeological material can be recorded during deep-ocean surveys, there may be no mechanism for proper recording and reporting in certain national or international waters and there is no central, internationally accessible repository where this data can be stored. Oil and gas companies in the US, for example, gather substantial amounts of information on shipwrecks because of the surveys they are required to conduct prior to approval of their permit. However, in other countries or in international waters, there might not be a legal obligation to report archaeological findings or make the data available to archaeologists.

The following case studies represent a sampling of the work that has been done at deep-water sites, not necessarily UCH that is at risk from DSM. The work has been conducted in disparate corners of the ocean and has employed a variety of groundbreaking techniques and methodologies. Some have captured the attention of the public and others the minds of other maritime archaeologists. Many also show the significance of shipwrecks to seafloor ecology and the potential collaboration between maritime archaeologists and marine biologists that can, and should, occur.

5.2 Case Studies

5.2.1 *Finding* Endurance

In March 2022, 107 years after the vessel sank, a team of scientists found *Endurance* in the Weddell Sea (Amos, 2022). *Endurance* was the vessel of Sir Ernest Shackleton's Antarctic expedition which was crushed by sea ice and sank in 1915.[2] The vessel was found more than 4 nautical miles from the sinking location noted by the Captain Frank Worsley and lay in roughly 3000 m of water. Thanks to this immense depth and the water conditions, the wreck is almost perfectly preserved: the name plate is still visible. The project used hybrid vehicles, which combined ROV and AUV features to search for the shipwreck. The AUV, equipped with a laser scanner and cameras, produced a 3D photogrammetric model of the wreck. Since its discovery, the wreck has been designated as a monument under the international Antarctic Treaty. The UK Antarctic Heritage Trust prepped a management plan for the wreck site in partnership with Historic England (Meara, 2024). To do so, the team created 'a detailed statement of significance, outlining why the site is important, and identifying the cultural and natural heritage values that need protection' (Meara, 2024).

This was an interdisciplinary project, and maritime archaeology was not the only focus. Onboard were sea ice scientists, oceanographers, meteorologists, and marine engineers to research the Antarctic environment. The project and discovery were well publicised and captured public attention around the globe. This even included a crowd-source project online where contributors identified deep-sea creatures visible in the wreck photos (McGreevy, 2022). Although the wreck is protected and not at risk from DSM, the project shows an example of effective wreck protection and management plans that focus on both natural and cultural heritage. UCH threatened by DSM could be protected with the same things considered.

[2] Project website: Falklands Maritime Heritage Trust. (2024). *Endurance 22 Expedition*. https://endurance22.org/.

5.2.2 *Black Sea MAP*

The Black Sea MAP (Maritime Archaeological Project) was based at Southampton University and conducted from 2015 to 2017. The goal was to study the waters of the Bulgarian Black Sea, where thousands of years ago, land was flooded as water levels rose after the last ice age.[3] The project involved geophysical surveys to detect the former land surfaces, collect samples, and answer key research question but it also had major finds maritime archaeology importance. Sixty-five shipwrecks were discovered, dating from the fourth century BC to the nineteenth century AD (Pacheco-Ruiz et al., 2019). Many of these wrecks were the first examples of certain ship types seen physically by archaeologists. Due to the anoxic conditions present in the Black Sea below 150 m, the wrecks are well preserved. The wrecks were located at depths ranging from 40 to 2200 m and the project shows 'the value of well-funded collaboration between academia and industry and that deep water archaeology can and must be totally in accordance to the 2011 United Nations Educational, Scientific and Cultural Organization (UNESCO) convention' (Pacheco-Ruiz et al., 2019, p. 1).

5.2.3 *Gulf of Mexico*

The Gulf of Mexico has been the primary source of oil and gas exploration and development in US waters since the 1940s (Priest, 2022). This expanded development has led to the discovery of hundreds of shipwrecks and other types of UCH since the 1970s. In 2004, the former US Minerals Management Service (MMS) (now Bureau of Ocean Energy Management or BOEM), partnered with NOAA's Office of Ocean Exploration to fund the Deep Gulf Wrecks Project. This project investigated seven World War II-era deep-water shipwrecks to study them archaeologically and to learn more about the fish and corals that colonised them to better understand the role of deep-water shipwrecks as artificial reefs (Church et al., 2007).

Other projects funded by MMS/BOEM have examined wrecks that were damaged by oil and gas industry activities. A 2003 deep-tow survey for a pipeline in the Viosca Knoll lease area detected a shipwreck at more than 600 m near the proposed pipeline area in the northeastern part of the Gulf (Church & Warren, 2008). MMS sponsored a site investigation in 2006, and the wreck was visually inspected by archaeologists with an ROV. The initial geophysical data suggested the vessel was a three-masted schooner, but the later ROV investigation showed it to be a possible nineteenth-century brig (Church & Warren, 2008). The side scan sonar data showed artefact scatter and numerous drag scars, determined to have been caused by large

[3] *Black Sea MAP | Centre for Maritime Archaeology | University of Southampton*. (n.d.). Retrieved September 2, 2024, from https://www.southampton.ac.uk/research/institutes-centres/centre-for-maritime-archaeology/black-sea-map.

anchor chain and cable associated with construction of the nearby oil and gas platform (Church & Warren, 2008). The location of the drag scars on the seabed correspond to a tear across the hull of the vessel. Archaeologists concluded that a buffer zone around the vessel would be sufficient protection for the wreck going forward (Church & Warren, 2008). The results of this archaeological analysis underscore the need for further investigation from an initial survey to better characterise the cultural heritage at risk.

One of the most cited examples of a wreck in the Gulf of Mexico is the 'Mardi Gras Shipwreck'. Sometime after 1815, the vessel sank in 1220 m of water off the coast of Louisiana and was discovered in 2002 by a pipeline survey for the Okeanos Gas Gathering Company (Ford et al., 2008). The wreck was spotted on video monitors from an ROV and in 2007 archaeologists, led by Texas A&M University, launched the deepest underwater excavation (at the time) to study the shipwreck and recover artefacts. The condition of the artefacts indicated that 'the low-energy environment of the deep Gulf is beneficial to the survival of artifacts', though iron objects and the vessel's hull showed more deterioration (Ford et al., 2008, p. v). The excavation is a key case study in deep-water archaeology as it 'allowed for the refinement of deep-water archaeological excavation methods' and was one of the only cases, worldwide, where an excavation was attempted at depth using an ROV (Ford et al., 2008, p. v). The archaeologists did, however, conclude that more refinement is needed, and better photogrammetry and excavation techniques must be developed through future deep-water excavations (Ford et al., 2008).

The use of telepresence has also been key to wreck investigations in the Gulf. NOAA's *Okeanos Explorer* visited Wreck 15563 in February 2022 and BOEM and other archaeologists and scientists were able to participate remotely to document the wreck during the mapping and ROV shakedown cruise the ship was conducting (Delgado & Brennan, 2023). The site was video documented with the ROV, and the onshore team examined the wreck, which lay at 1691 m off the coast of Alabama (Delgado & Brennan, 2023). The wreck was first discovered in 2011 during a side-scan sonar survey of an oil lease and reported to BOEM as a potential historic site (Delgado & Brennan, 2023). The 2022 mission emphasised 'the potential of telepresence for multiple, and multidisciplinary participation as well as reaching a wide and diverse audience' (Delgado & Brennan, 2023, p. 481).

The Gulf of Mexico is an ideal water body to study the long-term ecological role of wrecks and other human-made structures on the seafloor. More than 50 wrecks related to World War II lie in the Gulf and have been shown to serve as important artificial reefs (Church et al., 2007). In addition, BOEM has also been studying wooden wrecks in the deep waters of the Gulf and how they 'form islands of biodiversity on the seabed, shaping microbial ecology and ecosystem processes' (Hampel et al., 2022, p. 482). One of the wrecks, located at 1800 m, seems to be a key source of organic material at that depth, acting as an 'oasis in the desert' for 'microorganisms and a microbiome that is biologically distinct from the surrounding seabed of the Gulf of Mexico' (BOEM, 2022).

5.3 Areas Beyond National Jurisdiction

UCH does not exist only within territorial waters. *Titanic* is perhaps the most famous example of Ocean Heritage in international waters. The US Congress passed the 1986 R.M.S. *Titanic* Memorial Act after its discovery and, as Ballard argued, recommended the site be untouched as a memorial (Delgado, 2012). But, since the wreck is in The Area (the high seas outside of any nation's jurisdiction) it was accessible to anyone with the ability to reach it, so NOAA was also tasked with negotiating an international agreement for its protection and developing guidelines for activities on the wreck. This took more than a decade, and in the meantime, a private American company began diving on the wreck with IFREMER and controversially recovered more than 5000 artefacts (Delgado, 2012). The legal status of the wreck is complicated (see Aznar & Varmer, 2013) with several pieces of legislation protecting the wreck, including the 2000 Agreement Concerning the Shipwrecked Vessel R.M.S. *Titanic* negotiated by the UK, France, Canada, and the US; the Protection of Wrecks (R.M.S. *Titanic*) Order 2003 in the UK; and the US draft R.M.S. *Titanic* Maritime Memorial Preservation Act of 2009.

Titanic has certain protections in place but given the vast amount of ocean that is unmapped and un-surveyed for UCH, many sites do not have similar protection. It is precisely this UCH which is most at risk from seabed mining if not properly surveyed and protected. Logistics of excavations are much more difficult in international waters, so there are fewer case studies to use as examples of Ocean Heritage at risk. However, the lack of fieldwork does not mean that important heritage resources do not exist. A review of the historical context of the Emperor Seamounts is one example of critical UCH which lies in international waters, nonregulated by national jurisdiction (Delgado et al., 2024). These seamounts are submerged volcanic guyots with summit depths ranging from 250 to 1000 m and cited as ecologically and biologically significant by the Conference of the Parties to the Convention of Biological Diversity (Delgado et al., 2024). They lie near a maritime highway in the North Pacific which over millennia saw 'indigenous craft close to shore, Asiatic seafarers, and later vessels of colonial explorers, companies, and in time, the naval and merchant ships of modern nations and empires of the nineteenth and twentieth centuries' (Delgado et al., 2024, p. 42).

The Salas y Gómez and Nazca ridges are also examples of rich, culturally diverse Ocean Heritage in the areas beyond national jurisdiction (Delgado et al., 2022). This 'ranges from indigenous cultures who first ventured to this remote region close to a thousand years ago to the period of European colonial exploration, as well as the rise of the modern global economy' (Delgado et al., 2022, p. 1). The location also serves as a biodiversity hot spot.

5.4 Damage from Offshore Dredging

An existing industry with documented damage to Ocean Heritage that may be akin to DSM is offshore dredging. In the United States, studies have been done to look at the damage that the industry causes to archaeological sites (Research Planning Inc. et al., 2004). BOEM has jurisdiction over the mineral resources on the outer continental shelf and that 'includes the rights over sand, shell, or gravel resources for shore protection beach or wetland restoration projects, or for use in construction projects funded in whole or in part as authorized by the Federal Government' (Research Planning Inc. et al., 2004, p. 1). The US National Historic Preservation Act of 1966 requires federal agencies to protect cultural resources and so BOEM must do so with any dredging activity. BOEM (then MMS) commissioned a study to review its practices which found both direct and indirect impacts on submerged cultural resources. Direct impacts included those related to the direct dredge operations as well as those from ground tackle associated with mooring or manoeuvring the dredge. Indirect impacts came from exposure or burial of the cultural heritage. In other dredging cases, such as the shipwreck *Fame* in Sydney Harbour (Nutley, 1999) or *Le Juste* in the Loire River in France (Blot, 1996), entire shipwrecks have been destroyed, while other cases show less extreme damage, as is the case of the Averio Harbor wreck in Portugal (Alves, 1998; Blot, 2002). Some potentially polluting wrecks (see Chap. 6, this vol.) have also been observed with damage due to dredging activities including *Coimbra* (Brennan et al., 2023).

5.5 Steps Forward

Maritime archaeologists have an opportunity to lean into the brilliant work that has been conducted in the deep ocean. The visual media produced from surveys and excavations, including photo, video, 3D photogrammetry models, and more, should be used to grab the attention of the public and help debunk the myth of the dark, mysterious deep sea. Additionally, there is an opportunity for greater collaboration between marine sciences. This was well executed in the *Endeavor* discovery project and should continue. With the UN Decade of Ocean Science for Sustainable Development (2021–2030) pushing maritime research, the time is now for interdisciplinary projects. Interdisciplinary research is possible with the use of telepresence and live sharing of data. Embracing this is vital to deep-sea operations and it must be included in any and all operations. It also enables outreach to the general public, giving them a look into the field of archaeology and maritime sciences (Brennan et al., 2018).

Most importantly, there needs to be advocation for proper, scientific baseline surveys of the areas proposed for DSM and importantly, to have those surveys include archaeological analysis and reporting on the data. Archaeologists need to be brought in at the beginning of projects to make the operations easier and safer for all involved. Wrecks and other sites of Ocean Heritage can be avoided with proper surveys, interpretation, and archaeological reporting. These surveys need to be conducted, even in areas where some data already exists. Just because there has been work before does not mean the data gathered is of high enough quality to protect Ocean Heritage from mining activities. The lower resolution surveys and data processing that has been conducted in the past for deeper waters is often not sufficient for identifying UCH. Establishing regulations for DSM must include archaeologists, geologists, ecologists, and other scientists and stakeholders to make sure that the proper surveys, interpretations, exclusion zones, and mitigation measures are set.

Exclusion zones will be vital to protecting Ocean Heritage. For example, dredging in the United States requires a buffer zone around identified sites that varies from 30 m to 360 m (Research Planning Inc. et al., 2004). Other countries have different requirements for the size of the zone. Dredging companies were asked about the buffer zones in a 2004 survey, and many already had experience with them; exclusion zones have been used around 'hard-bottom sites', like coral reefs, for years (Research Planning Inc. et al., 2004). A one-size-fits-all approach should not be taken, however, as different sites have diverse conditions and impacts. Each site of Ocean Heritage 'should be investigated and the buffer zone determined' (Research Planning Inc. et al., 2004, p. 33) based on recommendations from archaeologists and environmental conditions. Monitoring must also take place where these zones are established from the advent of the activity, based on baseline data and surveys, and any changes must trigger a reconsideration of the zone (Research Planning Inc. et al., 2004).

Case Study Looking Forward: Archaeological Guidance

Mateusz Polakowski, MSDS Marine, Mat@msdsmarine.co.uk

Within the United Kingdom (UK) Exclusive Economic Zone (EEZ) all work that impacts the seabed must obtain a Marine Licence from the Marine Management Organisation (MMO) (Marine and Coastal Access Act, 2009). Part of the licencing process requires the MMO to consult with and obtain advice from several different organisations, including Historic England (HE). HE is a public body that is tasked with identifying and protecting heritage and archaeology on land and under the sea. In this capacity HE relies on archaeologists who can provide advice and strategies to determine what archaeology may lie on the seabed and the best methods for mitigation. The best tool for determining archaeological potential on and under the seabed is geophysical data (Side-Scan Sonar (SSS), Multibeam Echosounder (MBES), Magnetometer (MAG), and Sub Bottom Profiler (SBP) (Firth, 2010).

Geophysical survey (also sometimes referred to as remote sensing) is a non-destructive method of detecting, imaging, and studying seabed geology and submerged sites, including shipwrecks, paleolandscapes, and marine ecosystems.

Geophysical survey is the most cost effect way of carrying out assessments of the seabed. Marine geophysical data uses acoustics to image the seabed (SSS and MBES) or under the seabed (SBP), except for MAG which measures fluctuations in the Earth's magnetic field. This type of data, and its interpretation, is one of the core requirements of HE when providing comment and advice to the MMO on potential anthropogenic material within the marine environment.

The successful collection and archaeological interpretation of geophysical data is reliant on understanding the equipment, setting out clearly defined goals, and understanding the desired outcomes (i.e., what type of data you want to collect). It is worth noting at this point that the usual methodology is for an archaeological contractor to employ a hydrographic survey company to collect and process the data, with the archaeologists then undertaking the interpretation.

A guidance document like the *Marine Geophysics Data Acquisition, Processing and Interpretation (Second Edition)* (Historic England, 2025) provides archaeologists and developers within the UK with a framework for geophysical survey techniques so that survey campaigns can operate to specification standards necessary to detect the presence of historic and archaeological sites on or under the seabed. To ensure the most robust protection it is imperative that the geophysical data is viewed and interpreted by an archaeologist with a sound understanding of the data, the theory behind the data, the methods of collecting and positioning the data and how to use the software effectively to ensure that information is not lost, hidden or wrongly positioned.

This guidance was recently updated and redeveloped because it is essential that HE have an up-to-date guidance available that can explain how geophysical data can be used to reveal the presence of heritage assets so that development projects can be appropriately designed with mitigation measures which best support marine planning and national planning policy and objectives. The updated guidance ensures that fit for purpose, up to date evidence about the historic environment is gathered that can be used to make informed, proactive decisions about the marine historic environment.

The interpretation of geophysical and hydrographic data is, by its very nature, subjective. However, with experience and by analysing the form, size, and characteristics of an anomaly, a reasonable degree of certainty as to the origin of an anomaly can be achieved. There may be instances where an anomaly may exist on the seabed but not be visible in the geophysical data. This may be due to being covered by sediment or being obscured from the line of sight of the sonar. The combined use of various techniques like SSS and MBES data mitigates this by visualising anomalies from multiples angles, including from above. Where an anomaly is visible in the MBES data, that position will generally be used for the implementation of mitigation recommendations. The position obtained from the MBES data is generally more accurate due to the sensor and the GPS receiver being fixed to the vessel in known planes. SSS sensors are towed, and thus the margin for error is greater. The archaeological assessment of SBP data can used to corroborate sub-surface magnetic anomalies and identify the extents of larger surface anomalies such as partially buried wrecks.

Aggregate Extraction as an Example

Marine dredging and mineral extraction are not new developments. In the UK alone, the aggregate extraction industry has been active since the 1920s and a major contributor to the construction industry (Highley et al., 2007) and currently contributes 6.4% of the construction aggregates in Britain. Within this industry archaeologists regularly undertake assessments as part of the licensing process that identify, inform, and mitigate any detectable anthropogenic material on the seabed (BMAPA & English Heritage, 2003).

A standard guidance document, like the updated HE Geophysics Guidance, provides archaeologists with an important resource that can ensure regulatory compliance, clarity of requirements, consistency, and a shift towards greater efficiency, accountability, and mitigation of liability (Haeder & Yackee, 2020). This means that archaeological interpretation and mitigation efforts across aggregates extraction license areas can refer to this guidance as a means to ensure survey works are conducted to a specification suitable for archaeological analysis. This in turn ensures that the archaeological interpretation is carried out on data that is fit for purpose which enables the best possible identification, analysis, and mitigation efforts for anthropogenic material on or beneath the seabed.

References

Alves, F. (1998). A mid-15th century shipwreck from the west-central Portuguese coast. In *Pre-Proceedings of the International Symposium, "Archaeology of Medieval and Modern Ships of Iberian-Atlantic Traditions", Hull Remains, Manuscripts and Ethnographic Sources: A Comparative Approach* (pp. 63–71).

Amos, J. (2022, March 9). *Endurance: Shackleton's lost ship is found in Antarctic*. https://www.bbc.com/news/science-environment-60662541.

Aznar, M. J., & Varmer, O. (2013). The *Titanic* as underwater cultural heritage: Challenges to its legal international protection. *Ocean Development & International Law, 44*(1), 96–112. https://doi.org/10.1080/00908320.2013.750978

Beltrame, C., Costa, E., & Gay, G. (2022). Deep water archaeology in Italy and in the Tyrrhenian Sea. *Heritage, 5*(3), 2106–2122. https://doi.org/10.3390/heritage5030110. p. 2107.

Blot, J.-Y. (1996). *Underwater archaeology: Exploring the world beneath the sea*. Thames & Hudson.

Blot, J.-Y. (2002). New courses in Maritime archaeology on Portugal. In C. V. Ruppé & J. F. Barstad (Eds.), *International handbook of underwater archaeology* (pp. 465–495). Kluwer Academic/Plenum Press.

BOEM. (2022, March 9). *Shipwreck study unearths microbial diversity on seafloor | Bureau of Ocean Energy Management*. https://www.boem.gov/newsroom/ocean-science-news/shipwreck-study-unearths-microbial-diversity-seafloor.

Brennan, M. L., Delgado, J. P., Jozsef, A., Marx, D. E., & Bierwagen, M. (2023). Site formation processes and pollution risk mitigation of World War II oil tanker shipwrecks: Coimbra and Munger T. Ball. *Journal of Maritime Archaeology, 18*(2), 321–335. https://doi.org/10.1007/s11457-023-09365-4

Brennan, M. L., Cantelas, F., Elliott, K., Delgado, J. P., Bell, K. L. C., Coleman, D., Fundis, A., Irion, J., Van Tilburg, H. K., & Ballard, R. D. (2018). Telepresence-enabled maritime archaeological exploration in the deep. *Journal of Maritime Archaeology, 13*(2), 97–121. https://doi.org/10.1007/s11457-018-9197-z

Church, R., Warren, D., Cullimore, R., Johnston, L., Schroeder, W., Patterson, W., Shirley, T., Kilgour, M., Morris, N., & Moore, J. (2007). *Archaeological and biological analysis of World War II shipwrecks in the Gulf of Mexico: Artificial reef effect in deep water* (No. OCS Study MMS 2007-015). U.S. Dept. of the Interior, Minerals Management Service, Gulf of Mexico OCS Region. https://espis.boem.gov/final%20reports/4239.pdf.

Church, R. A., & Warren, D. J. (2008). *Viosca Knoll Wreck: Discovery and investigation of an early nineteenth-century wooden sailing vessel in 2,000 feet of water* (OCS Study No. MMS 2008-018). U.S. Department of the Interior, Minerals and Management Service. https://www.govinfo.gov/content/pkg/GOVPUB-I-95a10d76a3b1b17b6d064a63e10bb25d/pdf/GOVPUB-I-95a10d76a3b1b17b6d064a63e10bb25d.pdf.

Delgado, J. P. (Ed.). (1998). *Encyclopedia of underwater and maritime archaeology*. Yale University Press.

Delgado, J. (2012, June). Archaeology of Titanic. *Archaeology Magazine*. https://archaeology.org/issues/may-june-2012/features/archaeology-of-titanic/.

Delgado, J. P., Brennan, M. L., Rapu Haoa, S. A., Rapu Leong, J. H., Gaymer, C. F., Carabias, D., Stokes, E., & Wagner, D. (2022). The hidden landscape: Maritime cultural heritage of the Salas y Gómez and Nazca ridges with implications for conservation on the high seas. *Marine Policy, 136*, 104877. https://doi.org/10.1016/j.marpol.2021.104877

Delgado, J. P., & Brennan, M. L. (2023). Wreck 15563: An early nineteenth-century merchant vessel in the Gulf of Mexico. *Journal of Maritime Archaeology, 18*(3), 477–520. https://doi.org/10.1007/s11457-023-09370-7

Delgado, J. P., Brennan, M. L., Stokes, E., & Wagner, D. (2024). A forgotten maritime highway: Maritime cultural heritage of the emperor seamounts with implications for high seas conservation. *Journal of Maritime Archaeology, 19*(1), 41–80. https://doi.org/10.1007/s11457-024-09389-4

Ford, B., Borgens, A., Bryant, W., Marshall, D., Hitchcock, P., Arias, C., & Hamilton, D. (2008). *Archaeological Excavation of the Mardi Gras Shipwreck (16M01), Gulf of Mexico Continental Slope* (OCS Report No. MMS 2008-037).

Hampel, J. J., Moseley, R. D., Mugge, R. L., Ray, A., Damour, M., Jones, D., & Hamdan, L. J. (2022). Deep-sea wooden shipwrecks influence sediment microbiome diversity. *Limnology and Oceanography, 67*(2), 482–497. https://doi.org/10.1002/lno.12008

Krumholz, J. S., & Brennan, M. L. (2015). Fishing for common ground: Investigations of the impact of trawling on ancient shipwreck sites uncovers a potential for management synergy. *Marine Policy, 61*, 127–133. https://doi.org/10.1016/j.marpol.2015.07.009

McCann, A. M., & Oleson, J. P. (2004). Deep-water shipwrecks off Skerki Bank: The 1997 survey. *Journal of Roman Archaeology*.

McGreevy, N. (2022). Wreck of Shackleton's "Endurance" discovered in icy Antarctic depths. *Smithsonian Magazine*. Retrieved September 2, 2024, from https://www.smithsonianmag.com/smart-news/ernest-shackleton-ship-discovery-antarctic-explorer-history-180979702/.

Meara, H. (2024). Endurance Preserved: Safeguarding a shipwreck in the world's most inaccessible waters. *SeaVoice, 5*. https://www.seavoice.online/post/endurance-preserved-safeguarding-a-shipwreck-in-the-world-s-most-inaccessible-waters

Meyer-Kaiser, K. S., Mires, C. H., Kovacs, M., Kovacs, E., & Haskell, B. (2022). Structural factors driving benthic invertebrate community structure on historical shipwrecks in a large North Atlantic marine sanctuary. *Marine Pollution Bulletin, 178*, 113622. https://doi.org/10.1016/j.marpolbul.2022.113622

Nutley, D. (1999). *Report on the Fame rescue archaeological project*. Heritage Office.

Ødegård, Ø., Hansen, R. E., Singh, H., & Maarleveld, T. J. (2018). Archaeological use of Synthetic Aperture Sonar on deepwater wreck sites in Skagerrak. *Journal of Archaeological Science, 89*, 1–13. https://doi.org/10.1016/j.jas.2017.10.005

Pacheco-Ruiz, R., Adams, J., Pedrotti, F., Grant, M., Holmlund, J., & Bailey, C. (2019). Deep sea archaeological survey in the Black Sea – Robotic documentation of 2,500 years of human seafaring. *Deep Sea Research Part I: Oceanographic Research Papers, 152*, 103087. https://doi.org/10.1016/j.dsr.2019.103087

Paxton, A. B., McGonigle, C., Damour, M., Holly, G., Caporaso, A., Campbell, P. B., Meyer-Kaiser, K. S., Hamdan, L. J., Mires, C. H., & Taylor, J. C. (2024). Shipwreck ecology: Understanding the function and processes from microbes to megafauna. *BioScience, 0*, 1–13. https://doi.org/10.1093/biosci/biad084

Pearson, N., & Thompson, B. S. (2023). Saving two fish with one wreck: Maximizing synergies in marine biodiversity conservation and underwater cultural heritage protection. *Marine Policy, 152*, 105613. https://doi.org/10.1016/j.marpol.2023.105613

Pickens, C., Lily, H., Harrould-Kolieb, E., Blanchard, C., & Chakraborty, A. (2024). From what-if to what-now: Status of the deep-sea mining regulations and underlying drivers for outstanding issues. *Marine Policy, 169*, 105967. https://doi.org/10.1016/j.marpol.2023.105967

Priest, T. (2022). *History of the Gulf of Mexico offshore oil and gas industry during the deepwater era. Vol. 2: Shell Oil's deepwater mission to mars*. U.S. Department of the Interior, Bureau of Ocean Energy Management. 98 p. Contract No.: M08AX12677. Report No.: OCS Study BOEM 2022-077.

Research Planning, Inc., Tidewater Atlantic Research, Inc., & W.F. Baird & Associates Ltd. (2004). *Archaeological damage from offshore dredging: Recommendations for pre-operational surveys and mitigation during dredging to avoid adverse impacts* (OCS Study No. MMS 2004-005). Prepared for: U.S. Department of the Interior Minerals Management Service. https://www.govinfo.gov/content/pkg/GOVPUB-I-5810630d463e0d4a9861f9f74f2a66d2/pdf/GOVPUB-I-5810630d463e0d4a9861f9f74f2a66d2.pdf.

Søreide, F. (2011). *Ships from the depths: Deepwater archaeology* (1st ed.). Texas A&M University Press.

Appendix References

British Marine Aggregate Producers Association (BMAPA), & English Heritage. (2003). *Marine aggregate dredging and the historic environment guidance note*.

Firth, A. (2010). Marine geophysics: Integrated approaches to sensing the seabed. In D. Cowley (Ed.), *Remote Sensing for Archaeological Heritage Management: Proceedings of the 11th EAC Heritage Management Symposium, Reykjavik, Iceland, 25-27 March 2010* (pp. 129–140).

Haeder, S. F., & Yackee, S. W. (2020). Policies that bind? The use of Guidance Documents by Federal Agencies. *Journal of Health and Human Services Administration, 43*(2), 87–100. https://doi.org/10.37808/jhhsa.43.2.1

Highley, D. E., Hetherington, L. E., Brown, T. J., Harrison, D. J., & Jenkins, G. O. (2007). *The strategic importance of the marine aggregate industry to the UK* (No. Ref. OR/07/019; British Geological Survey Research Report).

Historic England. (2025). *Marine Geophysics: Data Acquisition, Processing and Interpretation Guidance Notes* (2nd Edition). Swindon. https://historicengland.org.uk/images-books/publications/marine-geophysics-data-acquisition-processing-interpretation/.

Marine and Coastal Access Act, 2009 c. 23. (2009). https://www.legislation.gov.uk/ukpga/2009/23.

Chapter 6
Possible Cumulative Impacts of Deep Sea Mining on World War II Potentially Polluting Wrecks

Robert S. Glover

Abstract As global demand for essential materials such as copper and manganese surges alongside the expansion of green energy and blue economy ventures, dwindling terrestrial supplies have made the deep seabed an increasingly attractive target for extracting these highly desirable metals. Advances in offshore exploration technology now make deep sea mining (DSM) a potential reality within the next decade, though its implications for underwater cultural heritage (UCH) remain largely unknown. In this chapter, the potential impacts of DSM on potentially polluting wrecks (PPWs)—World War II-era vessels that, laden with toxic cargoes, are beginning to leak harmful substances into the marine environment—will be explored. While the exact effects of mining activities near PPWs cannot be fully predicted until operations commence, speculative hypotheses can be developed by examining DSM's broader environmental impacts and reviewing case studies of similar phenomena affecting UCH and marine industries worldwide. Although DSM may not cause immediately noticeable changes to PPWs near mining sites, it could foster a marine environment increasingly hostile to submerged cultural assets. The resulting toxic mixture from mining activities and PPW leaks accelerated by mining may have far-reaching consequences for coastal communities and marine industries around the globe.

Keywords Deep sea mining · Potentially polluting wrecks · PPW · Pollution · Heritage · WWII

R. S. Glover (✉)
The Ocean Foundation, Oneonta, NY, USA

C. Jarvis (ed.), *Threats to Our Ocean Heritage: Deep Sea Mining*, SpringerBriefs in Archaeology, https://doi.org/10.1007/978-3-031-98238-5_6

6.1 Introduction

The remote and largely undisturbed environments that exist at the deepest points in the ocean can harbour some of the most complex ecosystems on the planet and increased exploration enables discoveries that continue to reshape our understanding of the natural world (Ramirez-Llodra et al., 2010; Przeslawski & Christenhusz, 2022; Danovaro et al., 2014). These locations typically experience minimal current movement, minimal oxygen content, low ambient temperatures, and low pH levels, contributing to stable seabed environments and a heightened capability to preserve underwater cultural heritage (UCH) (Luciano et al., 2013; MacLeod, 2002). The rapid growth of offshore exploration technologies has made both the natural and cultural resources of the deep sea more accessible to the scientific community, allowing more opportunities for discovery, research, and protection (Wenhai et al., 2019; Manglis et al., 2021; Brennan et al., 2018). With these advances, however, come new opportunities to exploit these remote areas, and the same technologies that serve as windows to this sensitive environment can create risks for both it and the fragile UCH found there (Washburn et al., 2019; Papageorgiou, 2019; Glover & Smith, 2003; Henderson, 2019).

In this chapter, the discussion will cover deep sea mining (DSM) and the adverse effects it could have on potentially polluting wrecks (PPWs), which is already an area of growing concern in the heritage and conservation community. As DSM operations are still in trial stages, the definitive effects that DSM could have on PPWs have yet to be determined. This chapter will suggest some potential effects DSM could have on PPWs by reviewing case studies, comparing similar phenomena in the heritage and salvage sectors, and relating applicable developments in blue sciences. It is important to note that the majority of listed potential effects will not directly or immediately cause the catastrophic release of harmful cargo into the water column but will instead contribute to the cumulative aggravation of the steel structures, encouraging expedited deterioration in conjunction with the vast array of pre-existing corrosive factors associated with steel shipwreck decay (Glover, 2024). While the likelihood of DSM impacting a PPW is not high, it is still important to draw attention to the threat that DSM could pose to UCH and the complicated toxic environment that PPWs and DSM could create for delicate deep-sea ecosystems.

6.2 Deep Sea Mining

Growing global demand for specific metals, like copper, manganese, nickel, and cobalt, has strained these materials' already over-exploited land-based deposits. An aggressive push for more readily available metals has led to a renewed interest in the deep-sea environment as a potential future source for them (Liu et al., 2023; Sparenberg, 2019). While previous attempts in the 1960s and 1970s to extract materials from the seabed were unsuccessful, recent technological growth has made

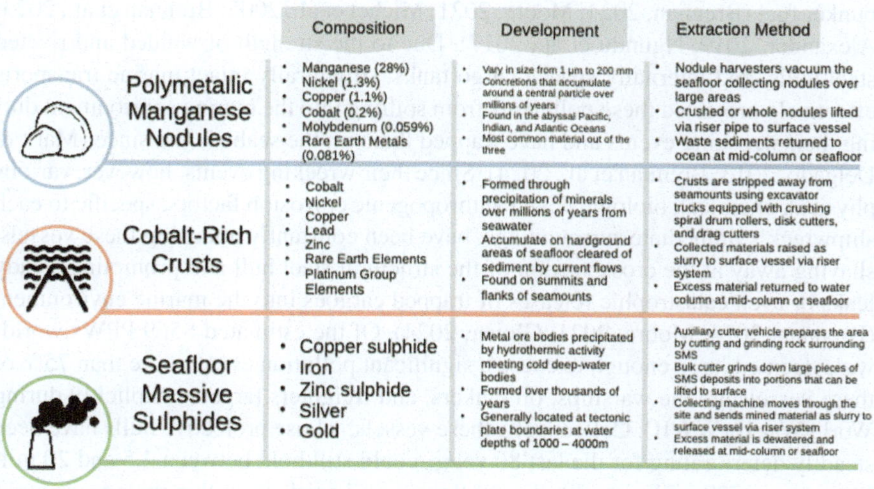

Fig. 6.1 Composition, Development, and Extraction Methods for the three primary materials targeted for DSM: Polymetallic Manganese Nodules (PMN), Cobalt-Rich Crusts (CRC), and Seafloor Massive Sulphides (SMS) (Sources: Levin et al., 2016; Liu et al., 2023. Image Sources: Maldjian, 2020; Oki Orlando, 2022; Wu, 2016)

DSM a more achievable option for resource accumulation (Sparenberg, 2019). Concerns regarding the general effects that DSM could have on the living environment have been well documented and have described wide-reaching, near-permanent impacts on deep-sea ecosystems (Miller et al., 2018; Kaikkonen et al., 2018; Cuyvers et al., 2018; Levin et al., 2016; Filho et al., 2021). These same effects could also potentially stress the delicate open systems that shipwrecks create within the marine environment (Quinn, 2006).

Currently, the International Seabed Authority (ISA) has approximately 30 mining exploration licenses in international waters for companies pursuing Polymetallic Manganese Nodules (PMN), Seafloor Massive Sulphides (SMS), and Cobalt-Rich Crusts (CRC) (Fig. 6.1). Some nations, like Papua New Guinea and Japan, have additionally shown interest in beginning mining activities within their Exclusive Economic Zones (EEZs) (Lin et al., 2019). While trial runs and sample excavations have been completed in some areas of the world, open mining has yet to be approved, as the ISA Regulations for DSM are not complete (Pickens et al., 2024). If DSM regulations are approved, however, serious consequences could threaten natural and cultural heritage within and near contract exploration zones.

6.3 Potentially Polluting Wrecks

A PPW is the wreck of a ship that was sunk while transporting significant quantities of hazardous cargo, which can include fuel oils and additives, lubricants, explosives, chemical weapons, and radioactive substances, as well as large amounts of bunker fuel (Brennan, 2024; Moore, 2021; Michel et al., 2005; Brennan et al., 2023; Alexander, 2019; Ngundu et al., 2017). Due to the strength of welded and riveted steel hulls, and internally spaced cargo tanks, structurally intact marine transports effectively protected these pollutants from spilling into the marine environment during their wrecking events and have trapped them on the seabed ever since (Marx & Delgado, 2013; Brennan et al., 2024). Since their wrecking events, however, various physical, chemical, biological, and anthropogenic corrosion factors, specific to each shipwreck's immediate surroundings, have been constantly attacking these vessels, shaving away at the cross-section of the structural steel hull and promoting steady leaks or even catastrophic releases of trapped cargoes into the marine environment (Brennan, 2024; Moore, 2021; Glover, 2024). Of the estimated 8569 PPWs worldwide deemed large enough to cause a significant pollution event, more than 75% of these vessels are the warships, oil tankers, and freighters targeted explicitly during World War II (WWII). Collectively, these vessels, whose protective hulls have been steadily deteriorating for the last 80 years, could still hold between 2.5 and 20 million tonnes (757 million to 6 billion gallons) of oil products within their cargo tanks and fuel bunkers (Michel et al., 2005; Landquist et al., 2017).

PPW risk assessment tools are essential to understanding the relative biological, economic, and social threats a spill from a PPW could pose to coastal communities. While many methods have been developed to assess these wrecks, most predict the effects of a hypothetical oil leak by evaluating sensitive assets located near the wreck, creating site-specific spill models, and calculating potential impacts, allowing for informed and prompt decision-making (Goodsir et al., 2019; NOAA, 2013a; Landquist et al., 2016; Alcaro et al., 2007; SPREP, 2002; Horn et al., 2024; Amir-Heidari et al., 2019; Zigic, 2004). Such assessments, however, typically require understanding the vessel's current location and condition to provide the most accurate timeline for and consequences of structural failure. Because most offshore archaeological surveys are concerned with wrecks found along coastal and continental margins, the vast majority of deep-water PPWs still need to be located or positively identified. Therefore, their positions in relation to sensitive coastal communities and DSM exploration zones have yet to be confirmed (Detlie, 2024; Keith & Evans, 2016). PPW spills in Ulithi Atoll, Micronesia, and near San Francisco, USA, from WWII-era wrecks have already shown the real-world destructive effects that even a fraction of a PPW's pollutants can have on shoreline ecosystems and fishing-dependent communities, as well as the need for spill preparation and impact assessment (LTC, 2006; Hill & Rodway-Dyer, 2024). PPW database and analysis projects in the past have steadily gathered the knowledge required to handle spills produced by historic wrecks, generating databases and best practices essential to their assessment and management (NOAA, 2013a; Goodsir et al., 2019; Landquist et al., 2016; Alcaro et al., 2007; SPREP, 2002). Without data on the exact locations of these wrecks, however, these projects are limited in their ability to predict the consequences of a potential spill accurately. As the International Union for the Conservation of Nature (IUCN) anticipates that 'marine pollution from sunken vessels will reach its highest level this decade,' it is clear that the path forward must include locating these vessels as soon as possible, assessing their conditions and polluting potentials, and, if necessary, proactively relieving them of their cargoes before they can catastrophically release (IUCN, 2023).

6.4 WWII Shipwreck Map

Currently, resources for approximating offshore deep-water WWII shipwreck locations are limited. The most essential, freely available resource for this task is the Esri ArcGIS mapping application 'Sunken Ships from the Second World War' (Heersink, 2022). This map was developed to sketch the locations of over 15,000 ships sunk during WWII using various publicly available hard copy and Internet sources. While some limitations in the data do exist, including a lack of data on ships sunk as a result of natural disasters or accidents, vessels moved following wrecking events, and precise geographic data for some wrecks, this application still stands as the most comprehensive resource available to aid in the hunt for lost WWII shipwrecks, providing a general starting point from which search operations can begin (Heersink, 2021).

To illustrate the potential impacts of DSM on PPWs, the locations of sunken WWII shipwrecks were transposed on a map of currently known contract exploration zones and seabed material distributions, gathered from work by Miller et al. (2018) and interpretations of the ISA contract zone maps (Miller et al., 2018:3; ISA, 2024) (Fig. 6.2). Economically viable deep-sea minerals outside of exploration zones have also been applied to the map to anticipate future mining operations closer to coastlines, provided the success of initial mining endeavours. Regionally and internationally recognised Marine Protected Area (MPA) data, provided by Protected Planet's Protected Areas Resource, has been added to illustrate the sensitive ecosystems that DSM and PPWs could potentially pollute (UNEP-WCMC & IUCN, 2024) (Fig. 6.11). Coastal and international fisheries and other areas frequented by large fishing vessels were also mapped by displaying the automatic identification system (AIS) location data of over 70,000 fishing vessels, as collected by Global Fishing Watch and analyzed by Kroodsma and colleagues (in dataset *fleet-daily-csvs-100-v2-2020*) (Kroodsma et al., 2018; Global Fishing Watch, 2021) (Fig. 6.12). Cataloguing periods of extended fishing time by larger vessels in 2020, the gradient map shows where fisheries or large schools of fish could be found internationally and was combined with the map of WWII wrecks to show the vulnerabilities that the fishing industry could face from a spill created by a PPW in deeper waters (Kroodsma et al., 2018; Global Fishing Watch, 2021). Both the MPA and fishery maps can be found later in the chapter.

An outline of coastal oiling data would require a more robust dataset and processing system, which were not available to the author during the creation of this chapter. Well-developed models of coastal oiling from PPWs can be found, however, on the Remediation of Underwater Legacy Underwater Threats (RULET) risk assessment database, which can provide an idea of how oil may be transported from submerged wrecks found within United States coastal waters and how ecological

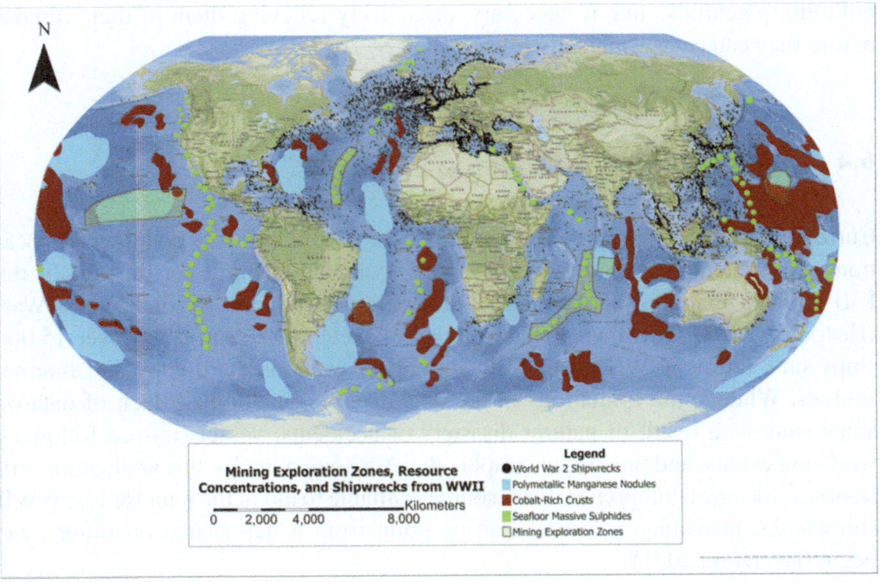

Fig. 6.2 The entire 'Sunken Ships from the Second World War' dataset transposed onto a map of approximate DSM contracted exploration zones and additional target materials. (Sources: Image made by author, 2024. Based on data from Heersink, 2022; Miller et al., 2018; ISA, 2024)

and anthropogenic communities found along American coastlines could be affected (NOAA, 2013b). The ultimate likelihood of oil from a deep-water shipwreck reaching coastlines, MPAs, or fisheries will be dictated by dynamic ocean processes, including, among other factors, surface and interfacial tension, surface currents, and wind waves. This is why predicting oil spread from a PPW can be so challenging (Ermakov et al., 2015).

6.4.1 Mid-Atlantic Ridge (Fig. 6.3)

The Mid-Atlantic Ridge potentially holds the largest number of WWII shipwrecks within its exploration zone, with 15 inside the zone and another 5 near its borders. Multiple tankers are on the list of vessels that could be lying in this zone, along with many cargo ships and several submarines, which may have been carrying various weapons or hazardous cargoes when they sank.

Name	Date	Country	Belligerent	Tonnage	Type	Cause
Frankfurt	08/04/41	Germany	Axis	5522	Cargo Ship	Scuttled
Everasma	02/28/42	Latvia	Allies	3644	Cargo Ship	Torpedo
Athelknight	05/27/42	UK	Allies	8940	Tanker	Torpedo
Patrick J Hurley	09/13/42	USA	Allies	10,865	Tanker	Torpedo
Holmpark	10/24/42	UK	Allies	5780	Cargo Ship	Torpedo
Etna	12/14/42	Sweden	Neutral	2619	Cargo Ship	Torpedo
British Vigilance	01/24/43	UK	Allies	8093	Tanker	Torpedo
Cape Decision	01/27/43	USA	Allies	5106	Cargo Ship	Torpedo
Rhexenor	02/03/43	UK	Allies	7957	Cargo Ship	Torpedo
St Margaret	02/27/43	UK	Allies	4312	Cargo Ship	Torpedo
<Null>	06/05/43	Germany	Axis	965	Submarine	Depth charge
Lot	06/22/43	France	Allies	4220	Tanker	Torpedo
<Null>	07/16/43	Germany	Axis	1120	Submarine	Depth charge
Henzada	07/24/43	UK	Allies	4161	Cargo Ship	Torpedo
Norderney	08/15/41	Germany	Axis	3667	Cargo Ship	Scuttled

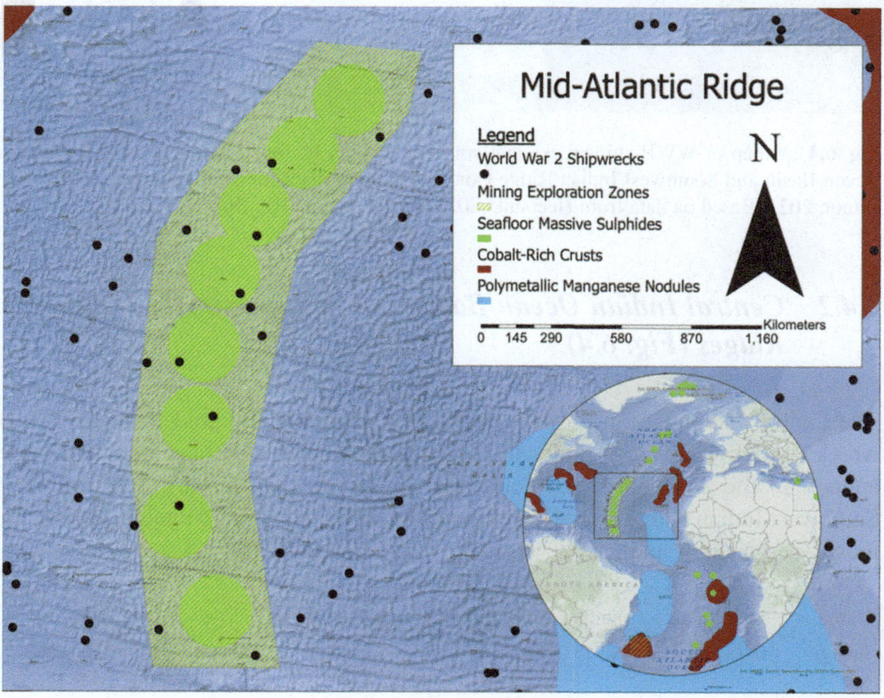

Fig. 6.3 A map of WWII shipwrecks that could be found within the Mid-Atlantic Ridge Contract Exploration Zone (Sources: Image made by author, 2024. Based on data from Heersink, 2022; Miller et al., 2018; ISA, 2024)

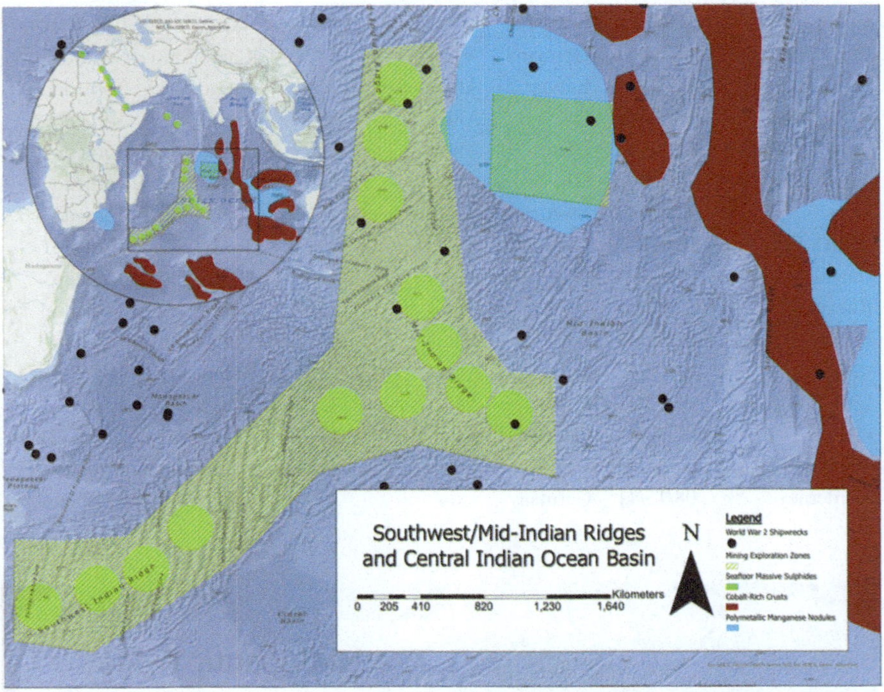

Fig. 6.4 A map of WWII shipwrecks that could potentially be found within the Central Indian Ocean Basin and Southwest Indian Ridge Contract Exploration Zones (Sources: Image made by author, 2024. Based on data from Heersink, 2022; Miller et al., 2018; ISA, 2024)

6.4.2 Central Indian Ocean Basin and Southwest/Mid-Indian Ridges (Fig. 6.4)

The Central Indian Ocean Basin exploration zone potentially holds just one WWII wreck, a UK-owned cargo ship called *Tulagi*. One other vessel can be found close enough to the zone border to potentially be affected by mining activities. The Southwest Indian Ridge and Mid-Indian Ridge Zone, however, contains six wrecks, mostly cargo ships, and another six close to its zone borders.

Name	Date	Country	Belligerent	Tonnage	Type	Cause
Tulagi	03/27/44	UK	Allies	2281	Cargo Ship	Torpedo

Name	Date	Country	Belligerent	Tonnage	Type	Cause
King City	08/24/40	UK	Allies	4744	Cargo Ship	Shelled
Athelking	09/09/40	UK	Allies	9557	Tanker	Shelled
Commissaire Ramel	09/20/40	France	Allies	10,001	Passenger Ship	Shelled
Benarty	09/10/40	UK	Allies	5800	Cargo Ship	Shelled
Fort Longueuil	09/19/43	UK	Allies	7128	Cargo Ship	Torpedo
Sutlej	02/26/44	UK	Allies	5187	Cargo Ship	Torpedo

6.4.3 Northwest Pacific (East Mariana Basin) (Fig. 6.5)

Five vessels, a mix of cargo ships and submarines, could be found in the Northwest Pacific exploration zone, with an additional six found near the borders.

Name	Date	Country	Belligerent	Tonnage	Type	Cause
Kaihei Maru	04/15/43	Japan	Axis	4575	Cargo Ship	Torpedo
Shoju Maru	08/05/43	Japan	Axis	1992	Cargo Ship	Torpedo
Soyo Maru	12/08/43	Japan	Axis	6081	Cargo Ship	Torpedo
<Null>	07/14/44	Japan	Axis	2564	Submarine	Depth charge
<Null>	06/27/45	Japan	Axis	1635	Submarine	Torpedo

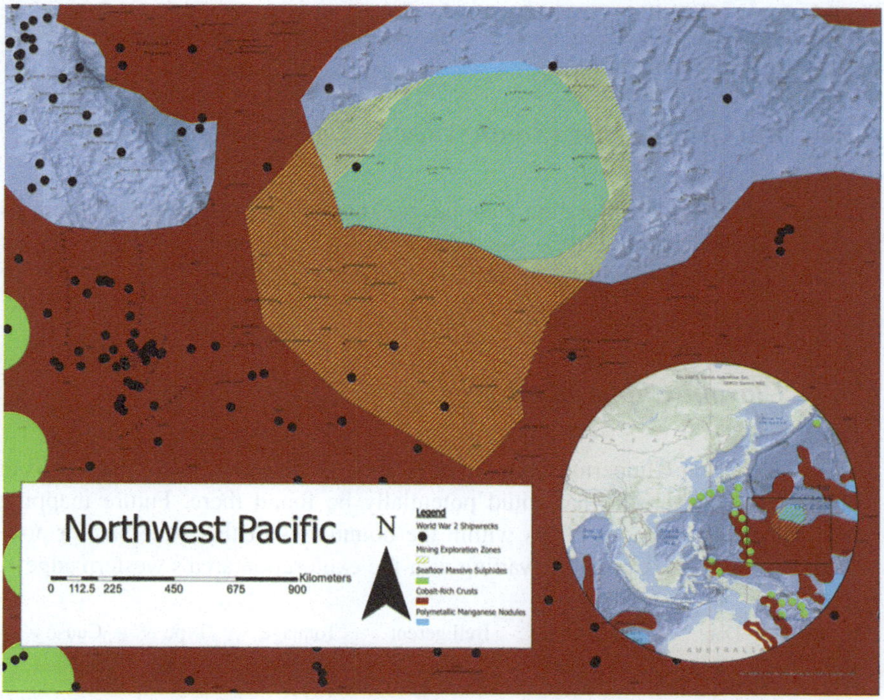

Fig. 6.5 A map of WWII shipwrecks that could potentially be found within the Northwest Pacific (East Mariana Basin) Contract Exploration Zone (Sources: Image made by author, 2024. Based on data from Heersink, 2022; Miller et al., 2018; ISA, 2024)

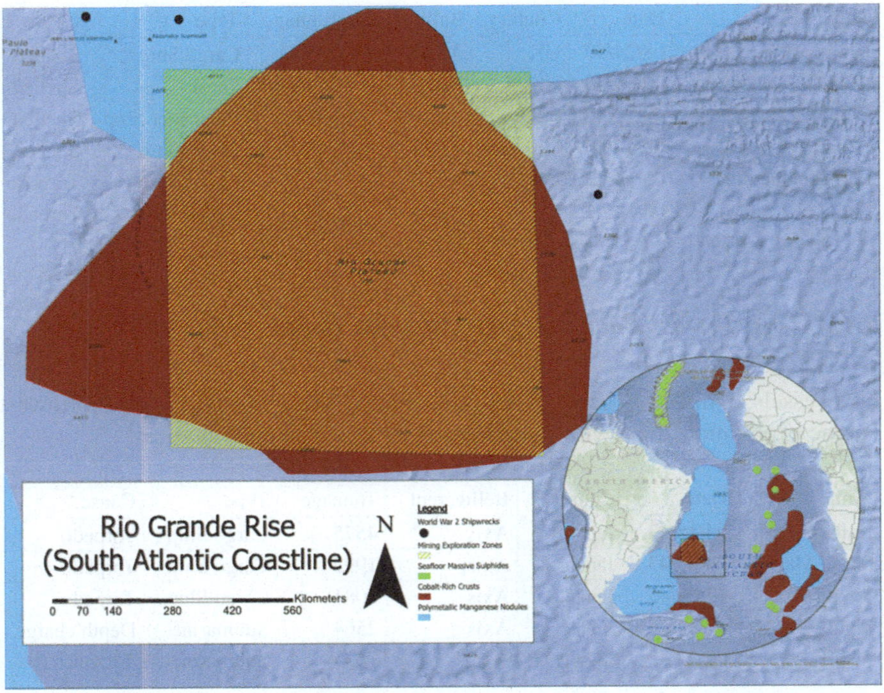

Fig. 6.6 A map of WWII shipwrecks that could potentially be found within the Rio Grande Rise Contract Exploration Zone (Sources: Image made by author, 2024. Based on data from Heersink, 2022; Miller et al., 2018; ISA, 2024)

6.4.4 Rio Grande Rise (South Atlantic Coastline) (Fig. 6.6)

The Rio Grande Rise exploration zone has no known WWII shipwrecks within its borders. Future mapping efforts could uncover a vessel lost within this zone, but currently, none are plotted here or are close to the zone borders.

6.4.5 Clarion-Clipperton Zone (Fig. 6.7)

Within the Clarion-Clipperton Zone, there is only one known vessel, a tanker from Panama called *Donerail*, that could potentially be found there. Future mapping efforts may uncover more PPWs within the boundaries of this zone but for now, there is just one wreck possibly waiting near the exploration area's western edge.

Name	Date	Country	Belligerent	Tonnage	Type	Cause
Donerail	12/09/41	Panama	Allies	4473	Tanker	Shelled

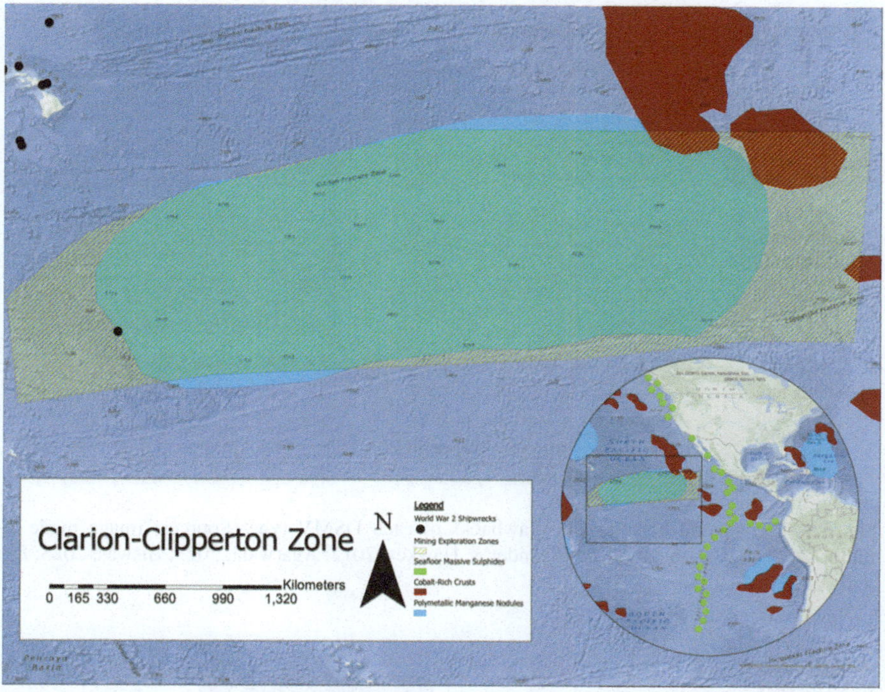

Fig. 6.7 A map of WWII shipwrecks that could potentially be found within the Clarion-Clipperton Contract Exploration Zone (Sources: Image made by author, 2024. Based on data from Heersink, 2022; Miller et al., 2018; ISA, 2024)

6.5 Mechanical Impact

A potential impact that DSM could have on PPWs would be a direct physical collision with the wreck from mining equipment. Deep sea mining vehicles (DSMVs) are crucial for extracting mineral resources from the seabed due to the extreme pressures experienced several kilometres below sea level and the subsequent difficult working conditions (Zhang et al., 2024). Successful mining vehicles are typically huge and densely packed with mining and sensory equipment, weighing several tons (Leng et al., 2021a; Mining.com, 2024; Liu et al., 2023; Hambling, 2018; Coffey Natural Systems, 2008). Heavily relied upon for all stages of mining operations, from seafloor mapping to full-scale mining operations, these vehicles generally fall into three categories: remotely operated vehicles (ROVs), autonomous underwater vehicles (AUVs), and hybrid remotely operated/autonomous underwater vehicles (HROVs) (Fig. 6.8).

A DSMV completing work near a PPW is a concern primarily due to the destructive potential associated with the mining vehicles' immense size. Over the last 30 years, at least 11 PMN collectors, 1 CRC extractor, and 3 SMS mining tools have reached trial stages. The PMN collectors have weighed on average, 15.29 tonnes;

DSMV Type	Advantages	Disadvantages
Remotely Operated Vehicle (ROV)	• Over 30+ years of use in offshore industry • Directly tethered to surface vessel, making vehicle loss difficult • Direct, precise control from highly trained operator in work activities • A wide range of tasks, both delicate and heavy, can be completed	• Limited speed capabilities during survey and cannot execute station-keeping activities in higher currents • Low visibility decreases effectiveness of video feedback sensors, causing production delays and increased costs • Require dedicated surface craft, increasing expenses • Human error is always a potential issue • Requires precise and dedicated control • Limited movement due to tether
Autonomous Underwater Vehicle (AUV)	• Extremely useful for broad area survey and large-scale, monotonous work • Increased speed capability and sophisticated navigation • Configurable, generally reliable, and versatile • Obtains high quality data • High level of positional accuracy and movement in low-transparency water using a wide array of oceanographic sensors • No umbilical tether • Advanced AI allows for more autonomy, including in identifying marine species	• Manual intervention required when at certain depths or in certain environmental conditions due to feedback delays • Has some difficulty breaking from pre-planned pathways • AI may have difficulty deciphering between separate marine targets • Trials to improve designs at depth can be expensive • Challenges in communication with machine • More support technology will be required to perfect use • Limited payload capacity • Requires significant amounts of power to operate
Hybrid Autonomous / Remotely Operated Vehicle (HROV)	• Combines advantages of both ROVs and AUVs • Can be directly tethered to vessel • Has overcome depth limits of 6,000 meters, a chokepoint for AUV development • Used for broad, automated tasks and more delicate, concentrated tasks • High level of positional accuracy from oceanographic sensors, with opportunity for manual operation in difficult navigational conditions	• Requires significant amounts of power to operate • Feedback delays from water depth and properties of deep-sea environment can create navigational problems • Human error is always a potential issue • Requires a dedicated surface craft and manual operator, increasing costs

Fig. 6.8 The general benefits and drawbacks of each DSMV type (Sources: Image made by author, 2024. Based on data from Elvander & Hawkes, 2012; Agarwala, 2023; Biswas, 2023; Liu et al., 2023)

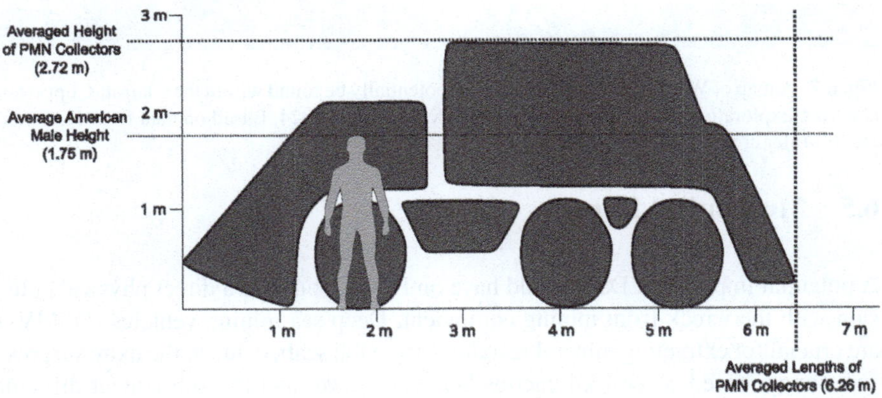

Fig. 6.9 The averaged height and length of all PMN collectors listed in publicly accessible sources compared to the height of an average American male, around 5′9″ or 1.75 m (Source: Image made by author, 2024. Based on data from Liu et al., 2023:1100–1102; Mining.com, 2024; Image Source: Dillon, 2020)

the CRC extractor weighed 6 tonnes; the SMS mining tools weighed between 200 and 300 tonnes (Liu et al., 2023:1100–1102; Mining.com, 2024; Hambling, 2018; Coffey Natural Systems, 2008:6) (Fig. 6.9). A single impact from a machine this large, moving at an assumed pace of 1–2 knots, could cause severe damages to the fragile corroded hull of a PPW which, to this point, has experienced several decades of submergence in the corrosive marine environment (Michel et al., 2005;

Abramowski et al., 2021). PPWs, such as *Coimbra* and SS *Coast Trader*, have in recent years been observed with significant damage and trapped cargo losses from impacts with scallop dredges, bottom trawling equipment, and other fishing gear (Brennan et al., 2023; Brennan, 2016; Delgado et al., 2018). Assuming impacts from standard fishing gear, averaging around a ton in weight and being pulled at approximately 1–2 knots, can cause this much damage to a PPW, more concern must be raised around the potential harm a self-powered DSMV could cause traveling at a similar speed (Alaska Department of Fish and Game, n.d.; Stokesbury et al., 2011; NOAA Fisheries, 2018; Demetri, 2016; He et al., 2021).

Deep-sea conditions are often extremely challenging to work in, and even slight delays in communications with a DSMV could result in hazardous consequences. A variety of obstacles, including ground deformation, slippage, hydrodynamic deep-sea current forces, riser suction strength, electromagnetic signal obstruction, and acoustic interference, can cause delays in equipment response, leading to real-time positioning failures and potential collisions with seabed obstacles (Liu et al., 2023). AUVs could encounter issues as they follow 2D pre-planned tracks that may not account for unanticipated changes, such as elevation or bedform changes or seabed obstacles, opening up the possibility for collisions or other accidents (Liu et al., 2023; Lu et al., 2023). Similarly, ROVs utilising forward-facing high-definition video for navigation could encounter challenges if their vision becomes clouded by sediment kickups generated during mining (Elvander & Hawkes, 2012). As even portions of PPWs, like the separated bows of SS *Bloody Marsh* and SS *Newhall Hills*, still have the potential to release oil despite their detachment from primary cargo tanks, large exclusion zones or complicated pathways would need to be created to avoid all portions of a wreck, which could be spread across an area of the seabed equal to 20% of the water's depth (Church, 2016; The Life-Boat, 1947; Brennan et al., 2024; iTVNews, 2022). A navigational complication due to environmental or mechanical problems, combined with a complex path around the site, could be concerning for mining operations near even a portion of a wreck. This could be further complicated by the loss of sea crawler control, which has occurred at least once during mining trials in the Pacific Clarion-Clipperton area in 2021. In this instance, while the collection device luckily remained immobile until a tether connection was established, future mishaps could be far less fortunate (Garanovic, 2021; Zhang et al., 2024).

Affording all PPWs a wide berth during planned mining operations could avoid unnecessary vessel risk and further expenses for mining groups. Currently, based on data drawn from the 'Sunken Ships from the Second World War' dataset, multiple tankers, cargo vessels, and submarines can be found in the mining exploration zones, potentially carrying various listed and unlisted cargoes (Heersink, 2022). There is always a chance that unreported or incorrectly recorded shipwrecks, both historic and more contemporary, could be found within or near DSM operations, still withholding toxic contents. The general ambiguity surrounding PPW cargo and wartime excess cargo not listed on shipping manifests could be harmful not just to the DSMVs themselves, which may not be equipped to interact with oil, chemical weapons, or potentially reactive explosive substances, but could also be dangerous to surface vessel crew members tasked with decontaminating the equipment after a PPW interaction (Chen et al., 2022; Pfeiffer, 2012; Christ & Wernli Sr., 2007; Brennan et al., 2023). Should a significant spill or catastrophic breach occur, costs would only increase for the already prohibitively expensive operations (Cardno, 2016; Sharma, 2011). Safety protocols would undoubtedly initiate a stop of work and actions would have to be taken to secure the slick and assess damages to nearby protected areas (API, 2013; UNEP-WCMC & IUCN, 2024). The remote locations of DSM operations, an assumed lack of oil spill equipment, and the continued costs of the support vessel and crew would drastically increase expenses (API, 2013; Schjølberg & Utne, 2015). Oil spill cleanup, which runs into the tens of thousands of USD per metric ton, would be added to reinvestment costs for deep-sea ecosystem restoration and long-term monitoring costs, already estimated to cost USD5.3–5.7 million per km^2 (Etkin, 1999; Moller et al., 1987; Amadi & Mosnier, 2023). While an extensive oil cleanup operation would eat into corporate profits, this could be avoided by removing fuel oils from the PPW before a spill could occur. Historically, these operations have averaged around USD23.9 million per PPW in more shallow waters, with deep-water wrecks relieved of oil via ROV, like *Prestige* and *Ehime Maru*, costing up to USD132 million (NOAA, 2013a). As it stands, however, this is still a manageable investment considering the significant initial capital required for mining, the cost-effectiveness of oil removal compared to spill cleanup, and the inevitable decrease in price that this operation will experience as technology, methodology, and availability improve (Loureiro et al., 2006; NOAA, 2013a; Lindgren et al., 2021; Laitenen, 2015).

6.6 Noise Pollution and Vibrations

Studies on the noise and vibrations generated by DSMVs and associated support vessels have indicated wide-reaching, severe impacts on marine ecosystems within the range of mining operations (Williams et al., 2022; Martin et al., 2021; Duarte et al., 2021; Mauro, 2020). The consequences that heightened sound propagation might have for UCH, however, have not received a significant amount of attention in the literature. For the last century, WWII PPWs found on the deep seabed have been deteriorating in an ocean growing steadily louder from low-frequency anthropogenic noises, such as commercial shipping and construction (Weilgart, 2017). While distant noise may not have created complications for the structural integrity of a PPW before, DSM would represent a much more direct and noticeable change in the wreck's immediate soundscape, potentially creating problems for the vessel's stability (Williams et al., 2022).

When DSM commences, noise above ambient levels will be generated throughout the water column, including at the surface (from acoustic exploration, including side-scan and multibeam sonar, and surface vessels), mid-column (from riser systems every ~1 km and submersibles), and seabed (from acoustic exploration, extraction vehicles, near-seafloor riser systems, and submersibles) (Martin et al., 2021;

Williams et al., 2022). About 61 times more energy is required for sound to transmit through water than through air, meaning that while sound in water is 61 dB quieter, energy will propagate further and maintain its pressure intensity more dependably than sounds on land (Webb, 2023). The typical ambient deep sea sound levels (~94 dB re 1 μPa) in the Pacific, for example, will be overpowered by the wide-reaching and intense noise generated by operational equipment (up to 195 dB re 1 μPa @ 1 m). The tight concentration of resources worldwide, and thus mining operations, would effectively envelop thousands to millions of km^2 of ocean floor in heightened noise emissions and associated vibrations (Williams et al., 2022; Hannington & Monecke, 2006; Secretariat for the International Seabed Authority Council, 2006; Levin et al., 2016). As most mining operations are expected to maintain regular to continuous operations up to or past two decades to maximise profitability, this would effectively transform these ecosystems into regions of near constant noise and vibration above ambient levels (Sharma, 2011; Office of Resources and Environmental Monitoring, 2006).

PPWs found in abyssal plains, where nodules are found, could suffer harmful vibrations from self-propelled crawlers, which dredge, scrape, and vacuum the seafloor during nodule collection. Drilling, scraping, grinding, and cutting completed by auxiliary and bulk cutters and crust excavator trucks, meanwhile, could cause structural weakness in vessels found near SMS and CRC mining zones, even potentially destabilising seamount flanks and causing landslides or collapses near a vessel (Liu et al., 2023; Hunt & Jarvis, 2020). While mining noise would probably not be strong enough to cause immediate concern to a PPW, continuous vibrations over extended periods could contribute to heightened vulnerability in damaged or already deteriorating locations around a historic vessel, creating a more substantial likelihood for collapse when combined with other pressures (Monti et al., 2013). More minor amplitude vibrations may also become dangerous when acting upon a structure that has already experienced a severe dynamic load, like traffic tremors acting upon a building affected by an earthquake or mining vibrations acting upon a vessel that suffered severe damages during a wrecking event or seabed impact (Monti et al., 2013; Kery & Stauffer, 2015).

Comparisons can be drawn to contextualise the effects of continuous or irregular vibrations on UCH near a mining site. Loud noises and vibrations generated within proximity to cultural material have been empirically linked to the loss of structural mass (Binal, 2014). Regulations for construction projects have become essential for historic building, cultural material, and museum artefact conservation due to the threats that impact generation can pose to historic sites and artefact integrity (Feilden, 2003; Xu et al., 2023; Zhu et al., 2023; Wei et al., 2018; Smyth et al., 2016; Johnson et al., 2013). Strong tremors, created by development activities, traffic, crowds, railways, and even loud concerts, have been connected to the structural damages of many historic buildings (Clemente & Rinaldis, 1998; Zini et al., 2021; Jones et al., 2011; Ma et al., 2016). Internal and external vibration reduction strategies have been intensely studied in offshore industries, such as oil rigs and pipelines, wind farms, and large shipping vessels, due to the long-term structural fatigue damage generated by regular and irregular shaking (Fa et al., 2020; Leng et al., 2021b;

Asiri & Al-Zahrani, 2014; Zhang et al., 2023; Xie et al., 2023; Victorovna, 2018; ABS, 2023). Guidelines for dredging operations near coastal structures and historic buildings have been imposed for many projects, as the vibrations they generate have been associated with facade and structural integrity damage (GHD, 2009; Ruggieri et al., 2024). Historic buildings, due to their fragility, require particularly conservative limits on tremors from offshore drilling and blasting, while pulsations from cruise ship engines have been identified as potential culprits for damages to Venice's historical foundations (Tripathy & Shirke, 2015; Bellingham et al., 2010). In a more extreme example, volcanic tremors in Indonesian waters shifted the USAT *Liberty* down a coastal slope, potentially causing abrasive damage to the ship's hull (Ridwan, 2019).

6.7 Sediment Displacement and Buildup

Sediment displaced because of mining activities could disrupt natural sediment deposition rates and become another potential contributor to the expedited breakdown of a PPW. Typically in the deep-sea environment, natural sediment deposition rates are negligible, averaging less than 1 cm per thousand years, and current velocity is predominantly too weak to resuspend bottom sediments, resulting in largely stable seabed conditions (Lyle, 2015; Levin et al., 2016; Keith & Evans, 2016; Chen et al., 2023) (Fig. 6.10). The potential for seabed scour or major sedimentation events for deep sea shipwrecks will typically occur during the initial deposition

Contract Exploration Zone	Sediment Type	Current Speeds
Clarion-Clipperton Zone	Varies by region but includes siliciclastic clay in north and central, biogenic calcareous oozes in south. Also includes radiolarian ooze and fine-grained calcareous sediment, with some reported gravel, coarser siliciclastic material and brown-clay sediments	Between 0.03 ms^{-1} and 0.25 ms^{-1}
Rio Grande Rise	Fine-grained carbonate ooze and red mud with Mn micronodules and nodules, with phosphatized FeMn crusts and phosphate-rich rocks along the top of the RGR plateau	Approximately 0.1 – 0.5 ms^{-1}
Northwest Pacific (East Mariana Basin)	Primarily biogenic carbonate sediments, with slightly carbonate argillaceous sediments and some limestones, breccias, and basalts	Approximately 3.6 ± 1.3 cms^{-1}
Southwest and Mid-Indian Ridges	Primarily calcium carbonate ooze, Fe-Mn(oxyhdr)oxide, and residual aluminosilicate phases, with some basaltic debris	Approximately 4.1±0.9 cm s^{-1}
Central Indian Ocean Basin	Siliceous oozes, red clay, and sporadic calcareous patches	Between <1 cm/s and 1-2 cm/s
Mid-Atlantic Ridge	Primarily pelagic carbonate ooze, which is thin or absent near the ridge but grows thicker as distance from ridge increases	Between 1-3 cm/s

Fig. 6.10 The general current structure and sediment makeup found within each of the contract exploration zones (Source: Image made by author, 2024. Based on data from Jones et al., 2021; Yanagimoto et al., 2010; Radziejewska et al., 2020; Vineesh et al., 2009; Sharma et al., 2020; Agarwal et al., 2020; Pollard & Read, 2017; ISA et al., 2020; Souza-Neto et al., 2024; Benites et al., 2021; Washburn et al., 2021)

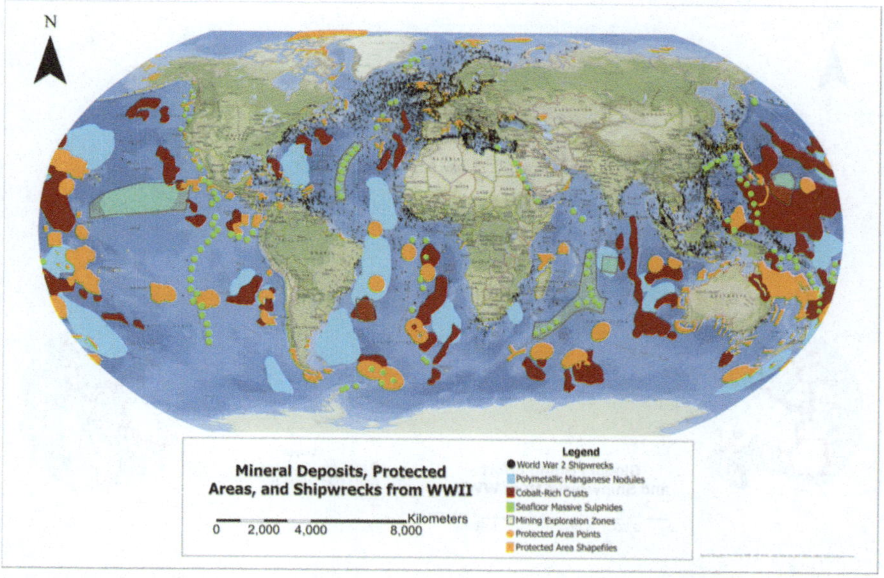

Fig. 6.11 DSM target materials and PPWs are close to many marine protected areas (MPAs) and important marine heritage sites worldwide. Hundreds of WWII wrecks that could be affected by DSM activities can be found near or on top of protected areas (Source: Sources: Image made by author, 2024. Based on data from UNEP-WCMC & IUCN, 2024; Heersink, 2022; Miller et al., 2018)

event or shortly thereafter, and the site will soon reach a state of stable disequilibrium within its surroundings, allowing for a steady decay rate (Quinn, 2006).

Significant changes in the marine landscape from DSM, however, and the associated shifts in sediment dynamics could contribute to considerable changes in the suspension and deposition of soft sediment found around a PPW, creating opportunities for the expedited breakdown of a ship (Levin et al., 2016). DSM could shift sediment dynamics in two significant ways: the direct resuspension of sediment and the alteration of deep-sea landforms. Sediment resuspension will inevitably result from seabed mining and will occur throughout mining operations. All DSMVs will create plumes of sediment as they work, whether that be as they dredge and plow through the top 15–40 cm of the seafloor to collect PMN, as they dig into the sides of seamounts to pry loose CRC, or as they grind down vent structures and remove sediment burden to collect SMS (Levin et al., 2016; Zhang et al., 2024). As the majority of deep-sea sediment types are pelagic clays, silts, deep sea oozes, and other cohesive sediments, the fine-grained material will remain suspended in the water column for longer than other sediment types and kickups could disperse via low-speed currents for up to 1400 km (Keith & Evans, 2016; Muñoz-Royo et al., 2021). Refuse sediment from ore processing reintroduced to the marine environment at the mid water column or directly at the seabed could provide another avenue for sediment suspension and dispersion, resulting in deposition far outside of the

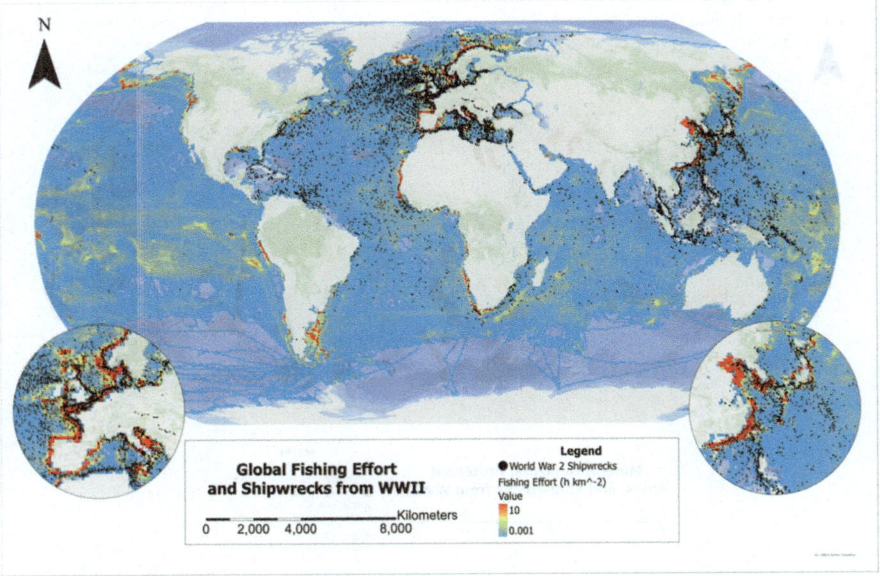

Fig. 6.12 PPWs can additionally be found near locations heavily trafficked and fished by trawlers, drifting longliners, and purse seiners, both in coastal and international waters. Should DSM contribute to the gradual leak or catastrophic release of a PPW's cargo, the resulting effects on fishing stock would be not just an environmental issue but could risk the livelihoods of artisan fishermen along coastlines or could harm the profits of larger fishing vessels working further offshore (Source: Sources: Image made by author, 2024. Based on data from et al., 2018; Global Fishing Watch, 2021)

initial mining area. Dispersion distance would depend on the remaining seafloor landscape, bottom current variability, sediment weight, deep-sea storm action, and other variables (Muñoz-Royo et al., 2021; Chen et al., 2023; Spearman et al., 2020).

The alteration of landforms and associated removal of hard substrate from the seabed could also affect the geochemical and hydrodynamic flow regimes, substratum characteristics, and sediment pathways of a mining location, creating new opportunities for soft sediment dispersal and sedimentation that a PPW may not have previously experienced (Levin et al., 2016; Van Dover, 2014; Wilkinson & Coroneos, 2017). Changes in flow regime would especially be seen as vertical topography in the marine environment is altered, when seamounts will be flattened to an extent, three-dimensional structures of hydrothermal vents and sulphide deposits will be levelled, and strip and open cast mining for polymetallic sulphides will be initiated (Levin et al., 2016; ISA, 2022). This could lead to the introduction of more soft sediment into already robust hydrodynamic regimes, like those found around seamounts and vents, leading to subsequent cumulative harm to biological or cultural materials found downstream from the mining (Levin et al., 2016). Seafloor topography could experience dramatic changes over the decades, especially considering the enormous size of mining locations. Removing hard substrate would thus allow large amounts of soft sediment to mobilise and spread across deep sea landscapes (Hannington & Monecke, 2006; Secretariat for the International Seabed Authority Council, 2006; Levin et al., 2016).

Sudden and extended resuspension of fine sediment particles in the deep-sea environment could have several effects on a PPW. While the weight of sediment underwater is lighter than in air due to the buoyant forces acting upon it, it is still creating new pressures upon a submerged steel structure that has been corroding for an extended period. Corrosion processes progressively thin mild steel by approximately ±0.1–0.4 mm per year, reducing structural mass and, therefore, structural strength. Much of the steel used for vessel construction during WWII ranged in thickness between 1 and 4 cm. After nearly 80 years of constant deterioration, their material structures are already drastically weakened, and any new weight could contribute to an expedited collapse (Glover, 2024). Sediment buildup is already calculated during marine salvage efforts due to the significant added strain it can create for both retrieval equipment and the wreck's hull, occasionally resulting in manual removal of the sediment burden (Ventikos et al., 2014). Similar to other rapidly degradable artefacts like wooden objects or fabrics that can be preserved by sediment cover, some ferrous artefacts too may experience lower degradation rates when buried in non-acidic soils conducive to the long-term preservation of iron artefacts. When introduced to an anaerobic environment under a sediment layer, however, many iron and steel objects may be targeted by sulphate-reducing bacteria that produce thick layers of corrosion products that trap an abundance of corrosive agents with the artefacts, accelerating breakdown (Dinh et al., 2004; Ashkenazi et al., 2017). An iron artefact with a portion of its length submerged beneath sediment, for example, may experience a much higher rate of corrosion than the portion that is raised above the seabed, as seen during studies of SS *Xantho* (MacLeod, 2016; McCarthy, 2000; MacLeod, 2002). This could potentially affect the decay rate of a ferrous-hulled wreck, portions of a wreck, or artefacts found within or on a wreck that are covered in sediment from mining operations, accelerating the timeline to failure.

Sediment movement across the hull could also create the potential for expedited corrosion. After submerging a vessel within the marine environment for an extended period, it will eventually be covered by concretions, or buildups of corrosion products, biofilms, sand particles, and static organisms (Melchers, 2003). Concretions serve as semi-permeable, anaerobic protective barriers between the bare metal and the seawater, stabilising the hull at a steady deterioration rate (North, 1976). Even particular biofilm layers have been observed protecting historic shipwrecks by forming protective coatings on shipwrecks, decreasing opportunities for abiotic corrosion (Mugge et al., 2019). If the concretion or biofilm layer is stripped from the hull due to the abrasive effects of altered flow regimes or resuspended sediments from mining and disposal, corrosion rates could increase as the metal becomes exposed again to the marine environment and vulnerable to rapid, targeted attacks (Moore, 2015; MacLeod et al., 2017; North & MacLeod, 1987). Variations in topography from seamounts and mid-ocean ridges can create strong, complex current flows, exhibiting significant variability and local intensity. Changes in these flow regimes could contribute to higher sediment-current velocity, more widespread dispersal of sediment following hard substrate removal, and potentially a more substantial possibility of concretion removal if a PPW is deposited downstream from mining operations (Hein & Petersen, 2013; Noble & Mullineaux, 1989; Lahaye et al., 2019). Even supposing a low current speed, the extended, decades-long operation periods for DSM operations could effectively 'sandblast' downstream PPWs with low amounts of fine-grained sediment for an extended period, potentially resulting in protective corrosion product and biolayer stripping, and an increased chance of hull deterioration (Levin et al., 2016; North & MacLeod, 1987).

6.8 Altered Corrosion Patterns

Rapid changes in the chemical makeup of waters surrounding a shipwreck could also influence the corrosive potential of steel. In the mid-water and seabed plumes used to return post-processed sediment and ore, trace amounts of heavy metals, including copper, manganese, nickel, cobalt, cadmium, zinc, and lead, can still be detected, which are then reintroduced into the water column from previous burial in sediment or ore (Hauton et al., 2017; Petersen et al., 2016; Levin et al., 2016). Heavy metals not only negatively influence biofilms found on the hull, damaging microbiomes that help protect and preserve it from abiotic corrosion, but they also have the potential to initiate galvanic coupling (de Araújo & de Oliveira, 2019; Mugge et al., 2019). Galvanic coupling occurs when metals of different corrosive potentials touch within an electrolytic solution, initiating heightened corrosion in the more reactive metal and lowered corrosion in the less reactive (MacLeod, 2019; North, 1984). Mild steel, the construction material used for most WWII ships, is more anodic, or reactive, than several of the heavy metals that will be released into the water column following processing, including copper, cadmium, lead, manganese, and nickel (Motion & Flow Control Products, Inc., n.d.; Hauton et al., 2017). While trace amounts of these materials will have no noticeable impact on the corrosion rates of the large hull plates, they may have more of an effect on hand-welded seams or the small rivets securing steel plating to the frame (North, 1984; Johnson et al., 2011; Masetti, 2012). If these trace amounts cover the structure of the vessel and form a galvanic couple with rivets, welds, or other structural fixtures, the galvanic reaction may reduce the components' ability to maintain the vessel's structural integrity, resulting in the detachment of plating and increased likelihood of corrosive attack (Brennan et al., 2023; MacLeod & Steyne, 2011). These trace metals could be carried, even by weak current action, deep into the vessel to create galvanic couplings, compromising its internal structure and damaging artifacts found there. Metals capable of initiating this reaction could also be deposited on or near wrecks as refuse lost from surface vessels and DSMVs (Arnshav, 2014; Sharma, 2020).

Oil released from the vessel's cargo tanks due to impacts from DSM or oil released from vessels involved in the mining could also cause damage to the PPW and affect its corrosion rate (Martin et al., 2021; Sharma, 2020). Oil introduced into the marine environment will typically float to the ocean surface and dissipate, evaporate, emulsify, or oxidate due to weathering processes (National Research Council, 2003; Horn et al., 2024). Some oil, however, may mix with suspended organic or inorganic matter in the water column, or even oil dispersants in the case of a response operation, and sink back to rest upon the hull as marine oiled snow sedimentation and flocculent accumulation (Hamdan et al., 2018; Paxton et al., 2024). Oil that resettles on the vessel can affect some of the micro- and macro-biota found on the hull, weakening the strength of biofilm communities that help to preserve steel from further corrosion (Mugge et al., 2019). Marine oiled snow sediment, as well as oil dispersants, are also known to encourage the presence of specific bacterial communities that can oxidise sulphur and produce caustic substances, while reducing their capacities to affect hydrocarbon degradation. This results in overall community composition shifts and increased corrosion rates, as seen on the German submarine *U-166* after being affected by the Deepwater Horizon spill in 2010 (Salerno et al., 2018; Mugge et al., 2019). If a spill from the surface vessel, mining equipment, or even the PPW itself sinks back onto the shipwreck, rates of steel breakdown could increase, leading to potentially more oil spills (Salerno et al., 2018; Birney et al., 2006; Sharma, 2020).

6.9 Archaeological Concerns

Deep sea wrecks can serve as living open-water museums, capable of providing important contextual data to aid in the interpretation of life aboard the vessel, acting as a time capsule of the ship's last working days, and, displaying, *in situ*, historic and unique vessels and technologies (Delgado & Varmer, 2015). Discovery, sometimes aided by offshore development industries, can be followed by the exploration of deep-sea wrecks via telepresence-enabled ROVs, allowing scientists and members of the public to directly engage with the wrecks (Brennan et al., 2018; Delgado et al., 2018; Church et al., 2009; Jones, 2023). While DSM surveys could aid in the future discovery of these ships, mining activities could also effectively restrict their study or mask them with sediment kickup. As of July 2025, the ISA has yet to finalise regulations on DSM, including those defining exclusion zones around a mining location (Northey, 2023). Moored and dynamically positioned oil rigs, however, have historically received 500 m radius exclusion zones, enforced by the United Nations Convention on the Law of the Sea (LOSC) and Coastal State legislation (Step Change in Safety, 2017; Napang et al., 2019). An exclusionary zone this large could impede research on identifying and assessing a PPW found within a mining zone, stifling opportunities to understand its historic nature and polluting potential (Lawrence et al., 2024). Sediment kickup from DSM could additionally restrict visibility for an ROV relying on high-definition video and potentially mask oil leaking from a ship to the ocean surface, a method that has been previously used to locate PPWs (Elvander & Hawkes, 2012; Brennan et al., 2024). Transparency of high-quality data on seabed obstructions and sufficient foresight in mining scheduling would allow for ROV exploration and study, PPW assessment, and even intervention measures to occur, if necessary, before mining proceeds.

Within archaeology, the context of artefacts, including their distribution and initial locations within the site matrix, is crucial in developing a more thorough interpretation of the site (Keith, 2016; Keith & Evans, 2016). As a vessel sinks, increasing internal pressures, mounting sinking speed, and internal and external drag will cause various portions of the vessel to break off and scatter, leaving a debris trail. This discard field illustrates the final moments of a vessel's working life. It can be used to track the movement of a ship during its wrecking process, aid in the search for shipwrecks from the same conflict, and dispel historical inaccuracies in battle logs (Church, 2016). Much of this crucial historical information could be crushed under, or further scattered by, the tracks of nodule collection vehicles, which typically crawl in a lawnmower pattern across the seafloor or could even be buried under landslide cover if deposited near a seamount destabilized by DSM (Liu et al., 2023; Hunt & Jarvis, 2020). Additionally, garbage, equipment lost overboard from the surface vessel, or pieces of equipment from DSMVs could harm the site's integrity, potentially introducing new galvanic coupling opportunities (Evans & Firth, 2016; Sharma, 2020).

6.10 Long-Term Effects

While the completion of DSM activities would abruptly remove the direct and proximate destructive influences from the immediate surroundings of the PPW, the biological and physical environment would be near-permanently altered, as all three seafloor deposit types require extensive periods to accumulate (Levin et al., 2016; Hein & Petersen, 2013). The direct alteration of the seafloor habitat, subsequent suspension of soft sediment, alteration of the substrate and its geochemistry, modification of sedimentation rates, and changes in substrate availability, heterogeneity, and flow regimes would continue to alter the corrosive environment of a PPW long after mining activities had concluded. The released heavy metals, chemicals, fuel contaminants, and the damage sustained from vibrations induced by the DSMV could also generate compounding structural instability within the wreck (Levin et al., 2016). Existing corrosion patterns would thus be permanently altered, potentially expediting corrosion rates and reducing the possibility of intervention before a catastrophic spill could occur.

The most significant long-term impact of successful DSM operations would be the continued growth of DSM. If the enormously expensive operations prove to be profitable, or if governments, for reasons other than economics, choose to subsidise them, then nations rich in seabed resources within their EEZs could initiate mining operations along their coastlines and continental shelves where the vast majority of shipwrecks are found (Gales, 2024; Cronan, 1990; Barrett, 2011). DSM would not only affect more densely concentrated numbers of shipwrecks along coastlines, but

PPWs deposited closer to shore would have a higher potential to oil coastlines, protected areas, and coastal communities, leading to more ecological, social, and economic damages (NOAA, 2013a; Saadoun, 2015; Andrews et al., 2021). As each nation will regulate mining within their own EEZ domestically using general international guidelines, environmental assessments and regulations may be weaker and compliance monitoring and enforcement may be more relaxed than under an impartial international authority. Currently, at least 13 nations have discussed opening deep sea mining operations within their EEZs, with a potential for that number to grow as the technology and funding becomes more readily available (Gales, 2024). While PPWs may have complicated meanings for different individuals and nations, they are still simultaneously the shared legacy of humanity and potentially harmful to the marine environment. Thus, they must be considered throughout all phases of material extraction (CHA, 2020; Jeffery, 2004).

The historic links that WWII wrecks serve as in modern society, connecting veterans, their families, and the general public to the memories of the lost, may be the strongest drivers for their continued identification and preservation. Despite their dangerous cargoes and the hazards they may pose to the marine environment, WWII wrecks still stand as tangible, albeit sometimes remote, historic sites connected to a significant period of international conflict (Brennan, 2024). Many serve as the final resting places for the thousands of sailors and civilians killed in wartime actions, and vessels more readily accessible to the public prompt reflection and education (Burlingame, 1991; Jeffery, 2004). For other groups, the wrecks symbolise both a painful period of significant change, marked by the loss of traditional ways of life and land during wartime, and a new source of income through tourism and heritage exploration (Jeffery, 2004). Some wrecks additionally derive their value from their proximity to other vessels, forming battleground memorials within nationally and internationally protected waters or delineating convoy routes and supply lines during more modern conflicts (Delgado & Varmer, 2015; Church, 2016; Hill et al., 2024). As many PPWs approach the 100-year threshold for international protection under the 2001 UNESCO Convention on the Protection of the Underwater Cultural Heritage, their continued preservation relies on persistent vigilance against possible threats and potentially harmful industries (Aznar & Varmer, 2024). Due to both Blue Economy growth and the heightened destructive potential of the marine environment from rapidly advancing climate change, shipwrecks face heightened daily threats (Mejjad & Rovere, 2021; Wright, 2016; Papageorgiou, 2019). Metallic wrecks continue to be harmed by development activities, despite protective measures and advanced detection technology, and even famous wrecks like *Titanic* are still subject to disturbance from recreational exploration (Evans & Firth, 2016; Research Planning, Inc. et al., 2004; Ballard & Sweeney, 2004). While DSM may not be as directly impactful to the structure of a vessel as steel salvage or dynamite fishing, it can still contribute to the increasingly hostile marine environment in which these vessels are found, potentially contributing to the collapse of the vessel, destruction of historic value, and release of oil (Evans & Firth, 2016; Browne, 2019; MacLeod et al., 2011; Barrett, 2023).

6.11 Conclusion

The potential effects that DSM could have on PPWs and the complicated impacts that both could inflict upon the marine environment must be further considered as momentum drives toward full-scale mining. While the intentions of DSM as an avenue towards new green energy opportunities are optimistic, they must come at a manageable cost to our marine environment and UCH. The deep-sea is a relatively untouched ecosystem and a living, breathing museum of submerged cultural heritage. DSM would alter this misunderstood treasure trove before studies could be completed, destroying locations of scientific intrigue before their impacts on planetary health can be fully explained. Caution must be maintained surrounding DSM, and the time must be taken to understand further what exists at the bottom of the ocean and how UCH fits into it before they both are permanently altered.

Acknowledgements I am extremely grateful to Paul Heersink for providing access to the data from the 'Sunken Ships of the Second World War' map. Mr. Heersink has dedicated countless hours to this invaluable project and has been extraordinarily generous in allowing me to use the resource for this chapter. This project would not have been possible without his commitment to this topic and willingness to share the knowledge he has gained to aid the research of others.

I also extend my heartfelt thanks to Charlotte Jarvis, Mateusz Polakowski, Benjamin Ferrari, Ole Varmer, and Chris Heeren for their insight, expertise, and support throughout the development of this chapter. Their thoughtful suggestions and feedback were invaluable as I attempted to navigate this unique topic. I am also profoundly grateful to The Ocean Foundation and Lloyd's Register Foundation for their generous support, which made it possible for me to dedicate the necessary time and resources to this work.

Finally, I am immensely appreciative of the countless individuals who reviewed and edited this chapter, offering both positive encouragement and helpful recommendations that greatly improved the overall content. Thank you all so much.

References

Abramowski, T., Urbanek, M., & Balaz, P. (2021). Structural economic assessment of polymetallic nodules mining project with updates to present market conditions. *Minerals, 11*(311), 1–22.

Agarwal, D. K., Roy, P., Prakash, L. S., & Kurian, P. J. (2020). Hydrothermal signatures in sediments from eastern Southwest Indian Ridge 63°E to 68°E. *Marine Chemistry, 218*, 1–16.

Agarwala, N. (2023). Using robotics to achieve ocean sustainability during the exploration phase of deep seabed mining. *Marine Technology Society Journal, 57*(1), 130–150.

Alaska Department of Fish and Game. (n.d.). *Weathervane Scallop*. Alaska Department of Fish and Game. Accessed August 29, 2024, from https://www.adfg.alaska.gov/index.cfm?adfg=weathervanescallop.uses.

Alcaro, L., Amato, E., Cabioch, F., Farchi, C., Gouriou, V., & Wrubl, C. (2007). *DEEPP Project: Development of European guidelines for Potentially Polluting Shipwrecks: Contract No. 0.7030900/2004/395842/SUB.A5*. Istituto centrale per la ricerca scientifica e tecnologica applicate al mare.

Alexander, D. E. (2019). The strange case of the *Richard Montgomery*: On the evolution of intractable risk. *Safety Science, 120*, 575–582.

Amadi, E., & Mosnier, F. (2023). *The sky high cost of deep sea mining*. Planet Tracker. Accessed October 26, 2024, from https://planet-tracker.org/wp-content/uploads/2023/06/Deep-Sea-Mining.pdf.

American Bureau of Shipping (ABS). (2023). *Guidance notes on ship vibration*. American Bureau of Shipping.

American Petroleum Institute (API). (2013). *Guidelines for offshore oil spill response plans: Guidance for offshore oil and gas exploration, production and pipeline facility operators*. API Technical Report 1145. API Publishing Services.

Amir-Heidari, P., Arneborg, L., Lindgren, J. F., Lindhe, A., Rosén, L., Raie, M., Axell, L., & Hassellöv, I.-M. (2019). A state-of-the-art model for spatial and stochastic oil spill risk assessment: A case study of oil spill from a shipwreck. *Environment International, 126*, 309–320.

Andrews, N., Bennett, N. J., Le Billon, P., Green, S. J., Cisneros-Montemayor, A. M., Amongin, S., Gray, N. J., & Rashid Sumaila, U. (2021). Oil, fisheries and coastal communities: A review of impacts on the environment, livelihoods, space and governance. *Energy Research and Social Science, 75*, 1–15.

Arnshav, M. (2014). The freedom of the seas: Untapping the archaeological potential of marine debris. *Journal of Maritime Archaeology, 9*, 1–25.

Ashkenazi, D., Nusbaum, I., Shacham-Diamand, Y., Cvikel, D., Kahanov, Y., & Inberg, A. (2017). A method of conserving ancient iron artefacts retrieved from shipwrecks using a combination of silane self-assembled monolayers and wax coating. *Corrosion Science, 123*, 88–102.

Asiri, S. A., & Al-Zahrani, Y. Z. (2014). Theoretical analysis of mechanical vibration for offshore platform structures. *World Journal of Mechanics, 4*, 1–11.

Aznar, M. J., & Varmer, O. (2024). Potentially polluting wrecks and the legal duty to protect our ocean heritage. In M. L. Brennan (Ed.), *Threats to our ocean heritage: Potentially polluting wrecks* (pp. 11–23). Springer.

Ballard, R. D., & Sweeney, M. (2004). *Return to Titanic*. National Geographic.

Barrett, M. J. (2011). *Potentially polluting shipwrecks: Spatial tools and analysis of WWII ship wrecks*. Master's project. Duke University.

Barrett, C. (2023, June 1). Chinese ship accused of looting iconic WWII shipwrecks. *Military Times*. Accessed October 27, 2024, from https://www.militarytimes.com/off-duty/military-culture/2023/06/01/chinese-ship-accused-of-looting-iconic-wwii-wrecks/.

Bellingham, S., Davis, C., O'Brien, C., & Saari, E. (2010). *Cruise control: To further the understanding of the comprehensive impacts of cruise ships on the city of Venice*. Worcester Polytechnic Institute.

Benites, M., Hein, J. R., Mizell, K., & Jovane, L. (2021). Miocene phosphatization of rocks from the summit of Rio Grande Rise, Southwest Atlantic Ocean. *Paleoceanography and Paleoclimatology, 36*(9), 1–24.

Binal, A. (2014). An investigation of noise effects on natural heritage and historic building materials. In G. Lollino, A. Manconi, F. Guzzetti, M. Culshaw, P. Bobrowsky, & F. Luino (Eds.), *Engineering geology for society and territory* (Vol. 5, pp. 1343–1346). Springer.

Birney, K., Griffin, A., Gwiazda, J., Kefauver, J., Nagai, T., & Varchol, D. (2006). *Potential deep-sea mining of seafloor massive sulfides: A case study in Papua New Guinea*. Donald Bren School of Environmental Science and Management. https://biblioteca.biofund.org.mz/wp-content/uploads/2018/11/1543243288-F1826.%20_Deep%20Sea%20Mining.Pdf.

Biswas, A. (2023). *AUV design and development for deep sea mining (polymetallic nodules)*. Maritime Research Center: Foundation for Underwater Domain Awareness.

Brennan, M. L. (2016). Quantifying impacts of trawling to shipwrecks. In M. E. Keith (Ed.), *Site formation processes of submerged shipwrecks* (pp. 157–178). University Press of Florida.

Brennan, M. L. (2024). Potentially polluting wrecks: An introduction. In M. L. Brennan (Ed.), *Threats to our ocean heritage: Potentially polluting wrecks* (pp. 1–10). Springer.

Brennan, M. L., Cantelas, F., Elliott, K., Delgado, J. P., Bell, K. L. C., Coleman, D., Fundis, A., Irion, J., Van Tillburg, H. K., & Ballard, R. D. (2018). Telepresence-enabled maritime archaeological exploration in the deep. *Journal of Maritime Archaeology, 13*, 97–121.

Brennan, M. L., Delgado, J. P., Jozsef, A., Marx, D. E., & Bierwagen, M. (2023). Site formation processes and pollution risk mitigation of World War II oil tanker shipwrecks: Coimbra and Munger T. Ball. *Journal of Maritime Archaeology, 18*(2), 321–355.

Brennan, M. L., Thiemann, G., & Jeffery, W. (2024). Satellite detection and the discovery of *Bloody Marsh*. In M. L. Brennan (Ed.), *Threats to our ocean heritage: Potentially polluting wrecks* (pp. 117–128). Springer.

Browne, K. (2019). "Ghost Battleships" of the Pacific: Metal pirates, WWII heritage, and environmental protection. *Journal of Maritime Archaeology, 14*(1), 1–28.

Burlingame, B. (1991). The USS Arizona memorial. *Naval History, 5*, 4. https://www.usni.org/magazines/naval-history-magazine/1991/december/uss-arizona-memorial

Cardno. (2016). *An assessment of the costs and benefits of mining deepsea minerals in the Pacific Island Region: Deep-sea mining cost-benefit analysis. SPC Technical Report SPC00035*. Pacific Community. Accessed August 29, 2024, from https://www.sprep.org/attachments/VirLib/Regional/deep-sea-mining-cba-PICs-2016.pdf

Chen, S.-Y. S., Ouillon, R., Muñoz-Royo, C., & Peacock, T. (2023). Oceanic bottom mixed layer in the Clarion-Clipperton Zone: Potential influence on deep-seabed mining plume dispersal. *Environmental Fluid Mechanics, 23*, 579–602.

Chen, X., Bose, N., Brito, M., Khan, F., Millar, G., Bulger, C., & Zou, T. (2022). Risk-based path planning for autonomous underwater vehicles in an oil spill environment. *Ocean Engineering, 266*(4), 1–13.

Christ, R. D., & Wernli, R. L., Sr. (2007). *The ROV manual: A user guide for observation-class remotely operated vehicles*. Elsevier.

Church, R. A. (2016). The U-166 and Robert E. Lee Battlefield: The equation of site distribution. In M. E. Keith (Ed.), *Site formation processes of submerged shipwrecks* (pp. 249–258). University Press of Florida.

Church, R. A., Warren, D. J., & Irion, J. B. (2009). Analysis of deepwater shipwrecks in the Gulf of Mexico: Artificial reef effect of six World War II shipwrecks. *Oceanography, 22*(2), 50–63.

Clemente, P., & Rinaldis, D. (1998). Protection of a monumental building against traffic-induced vibrations. *Soil Dynamics and Earthquake Engineering, 17*, 289–296.

Coffey Natural Systems. (2008). *Environmental impact statement – Nautilus Minerals Niugini Limited - Solwara 1 Project: Executive summary*. Coffey Natural Systems Pty Ltd.

Cronan, D. S. (1990). Overview of mineral resources in the EEZ. In D. A. Ardus & M. A. Champ (Eds.), *Ocean resources. Volume I: Assessment and utilisation* (pp. 105–111). Springer.

Cultural Heritage Agency: Ministry of Education, Culture, and Science (CHA). (2020). *Good practice: International collaboration on shared difficult heritage*. Cultural Heritage Agency: Ministry of Education, Culture, and Science. Accessed November 16, 2024, from https://english.cultureelerfgoed.nl/publications/publications/2020/01/01/good-practice-international-collaboration-on-shared-difficult-heritage.

Cuyvers, L., Berry, W., Gjerde, K., Thiele, T., & Wilhem, C. (2018). *Deep seabed mining: A rising environmental challenge*. International Union for Conservation of Nature and Natural Resources (IUCN).

Danovaro, R., Snelgrove, P. V. R., & Tyler, P. (2014). Challenging the paradigms of deep-sea ecology. *Trends in Ecology & Evolution, 29*(8), 465–475.

de Araújo, L. C. A., & de Oliveira, M. B. M. (2019). Effect of heavy metals on the biofilm formed by microorganisms from impacted aquatic environments. In S. Dincer, M. S. Özdenefe, & A. Arkut (Eds.), *Bacterial biofilms*. InTechOpen. https://www.intechopen.com/chapters/69536

Delgado, J. P., & Varmer, O. (2015). The public importance of World War I shipwrecks: Why a state should care and the challenges of protection. In U. Guérin, A. R. da Silva, & L. Simonds (Eds.), *Underwater cultural heritage from World War I: Proceedings of the Scientific Conference on the Occasion of the Centenary of World War I: Bruges, Belgium 26 & 27 June 2014* (pp. 105–116). UNESCO.

Delgado, J. P., Cantelas, F., Symons, L. C., Brennan, M. L., Sanders, R., Reger, E., Bergondo, D., Johnson, D. L., Marc, J., Schwemmer, R. V., Edgar, L., & MacLeod, D. (2018). Telepresence-enabled archaeological survey and identification of SS *Coast Trader*, Straits of Juan de Fuca, British Columbia, Canada. *Deep-Sea Research Part II, 150*, 22–29.

Demetri, J. (2016). *The evolution of New Bedford Fishing gear: Otter Trawling*. New Bedford Fishing Heritage Center. Accessed October 23, 2024, from https://fishingheritagecenter.org/wp-content/uploads/2016/04/Fishing-Vessel-Evolution-Otter-Trawl.pdf

Detlie, J. (2024). Searching for a lost PPW: SS *William Rockefeller*. In M. L. Brennan (Ed.), *Threats to our ocean heritage: Potentially polluting wrecks* (pp. 129–141). Springer.

Dinh, H. T., Kuever, J., Mußmann, M., Hassel, A. W., Stratmann, M., & Widdel, F. (2004). Iron corrosion by novel anaerobic microorganisms. *Nature, 427*, 829–832.

Duarte, C. M., Chapuis, L., Collin, S. P., Costa, D. P., Devassy, R. P., Eguiluz, V. M., Erbe, C., Gordon, T. C., Halpern, B. S., Harding, H. R., Havlik, M. N., Meekan, M., Merchant, N. D., Miksis-Olds, J. L., Parsons, M., Predragovic, M., Radford, A. N., Radford, C. A., Simpson, S. D., Slabbekoorn, H., Staaterman, E., Van Opzeeland, I. C., Winderen, J., Zhang, X., & Juanes, F. (2021). The soundscape of the Anthropocene ocean. *Science, 371*, 1–10.

Elvander, J., & Hawkes, G. (2012). *ROVs and AUVs in support of marine renewable technologies* (pp. 1–6). 2012 Oceans Conference, 14–19 October, 2012. Institute of Electrical and Electronic Engineers.

Ermakov, S., Kapustin, I., Sergievskaya, I., & da Silva, J. C. B. (2015). Spreading of oil films on the sea surface: Radar/optical observations and physical mechanisms. In *Conference on remote sensing of the ocean, sea ice, coastal waters, and large water regions* (pp. 1–8). SPIE.

Etkin, D. S. (1999). Estimating cleanup costs for oil spills. *International oil Spill Conference Proceedings, 1999*(1), 35–38.

Evans, A. M., & Firth, A. (2016). Anthropogenic impacts of development-led archaeology in an offshore context. In M. E. Keith (Ed.), *Site formation processes of submerged shipwrecks* (pp. 133–156). University Press of Florida.

Fa, Z., Xi, Q., Li, Y., Huang, H., & Rabczuk, T. (2020). Hybrid FEM–SBM solver for structural vibration induced underwater acoustic radiation in shallow marine environment. *Computer Methods in Applied Mechanics and Engineering, 369*, 1–23.

Feilden, B. M. (2003). *Conservation of historic buildings* (3rd ed.). Architectural Press.

Filho, W. L., Abubakar, I. R., Nunes, C., Platje, J., Ozuyar, P. G., Will, M., Nagy, G. J., Al-Amin, A. Q., Hunt, J. D., & Li, C. (2021). Deep seabed mining: A note on some potentials and risks to the sustainable mineral extraction from the oceans. *Journal of Marine Science and Engineering, 9*(5), 1–16.

Gales, P. (2024). 11 Countries developing subsea minerals in their EEZs. *Deep Sea Mining*. Accessed November 7, 2024, from https://deepseamining.ac/article/50#gsc.tab=0

Garanovic, A. (2021, April 29). Deep sea mining robot gets stuck 4,500 metres beneath Pacific Ocean's surface. *Offshore Energy*. https://www.offshore-energy.biz/deep-sea-mining-robot-gets-stuck-4500-metres-beneath-pacific-oceans-surface/#:~:text=Global%20Sea%20Mineral%20Resources%20(GSR,on%20the%20Pacific%20Ocean's%20seabed.

GHD. (2009). *Report for western basin dredging and disposal project: Noise and vibration assessment*. GHD.

Glover, R. (2024). Corrosion processes of steel-hulled potentially polluting wrecks. In M. L. Brennan (Ed.), *Threats to our ocean heritage: Potentially polluting wrecks* (pp. 41–59). Springer.

Glover, A. G., & Smith, C. R. (2003). The deep-sea floor ecosystem: Current status and prospects of anthropogenic change by the year 2025. *Environmental Conservation, 30*(3), 219–241.

Goodsir, F., Lonsdale, J. A., Mitchell, P. J., Suehring, R., Farcas, A., Whomersley, P., Brant, J. L., Clarke, C., Kirby, M. F., Skelhorn, M., & Hill, P. G. (2019). A standardized approach to the environmental risk assessment of potentially polluting wrecks. *Marine Pollution Bulletin, 142*, 290–302.

Hambling, D. (2018). *Giant robots are the future of underwater mining*. Popular Mechanics. Accessed October 20, 2024, from https://www.popularmechanics.com/technology/robots/a16674275/underwater-robot-mining-nautilus-solwara-1-papua-new-guinea/.

Hamdan, L. J., Salerno, J. L., Reed, A., Joye, S. B., & Damour, M. (2018). The impact of the *Deepwater Horizon* blowout on historic shipwreck-associated sediment microbiomes in the northern Gulf of Mexico. *Scientific Reports, 8*(9057), 1–14.

Hannington, M., & Monecke, T. (2006). *Global exploration models for polymetallic sulphide deposits in the area: Possible criteria for lease block selection under the draft regulations on prospecting and exploration for polymetallic sulphides*. Prepared for the International Seabed Authority. University of Ottawa.

Hauton, C., Brown, A., Thatje, S., Mestre, N. C., Bebianno, M. J., Martins, I., Bettencourt, R., Canals, M., Sanchez-Vidal, A., Shillito, B., Ravaux, J., Zbinden, M., Duperron, S., Mevenkamp, L., Vanreusel, A., Gambi, C., Dell'Anno, A., Danovaro, R., Gunn, V., & Weaver, P. (2017). Identifying toxic impacts of metals potentially released during deep-sea mining—A synthesis of the challenges to quantifying risk. *Frontiers of Marine Science, 4*, 1–13.

He, P., Chopin, F., Suuronen, P., Ferro, R. S. T., & Lansley, J. (2021). *Classification and illustrated definition of fishing gears*. FAO Fisheries and Aquaculture Technical Paper 672. Food and Agriculture Organization of the United Nations. https://openknowledge.fao.org/server/api/core/bitstreams/d194ba36-af7a-48ed-9e58-563143b96863/content.

Heersink, P. (2021, October 23). Project Scope. *Sunken ships of the Second World War*. Accessed October 27, 2024, from https://ww2sunkenships.ca/.

Hein, J. R., & Petersen, P. (2013). The geology of cobalt–rich ferromanganese crusts. In E. Baker & Y. Beaudoin (Eds.), *Deep sea minerals: Cobalt-rich ferromanganese crusts; A physical, biological, environmental, and technical review* (pp. 7–14). Secretariat of the Pacific Community.

Henderson, J. (2019). Oceans without history? Marine cultural heritage and the sustainable development agenda. *Sustainability, 11*(18), 1–22.

Hill, P. G., & Rodway-Dyer, S. (2024). Decolonising environmental risk assessments of Potentially Polluting Wrecks: A case study of the wreck of the USS Mississinewa in Ulithi Lagoon, Federated States of Micronesia. *Environmental Management, 73*(5), 973–984.

Hill, P. G., Skelhorn, M., & Goodsir, F. (2024). Managing potentially polluting wrecks in the United Kingdom. In M. L. Brennan (Ed.), *Threats to our ocean heritage: Potentially polluting wrecks* (pp. 71–83). Springer.

Horn, M., French-McCay, D., & Schmidt Etkin, D. (2024). Environmental impact and modeling of petroleum spills. In M. L. Brennan (Ed.), *Threats to our ocean heritage: Potentially polluting wrecks* (pp. 25–39). Springer.

Hunt, J. E., & Jarvis, I. (2020). The lifecycle of mid-ocean ridge seamounts and their prodigious flank collapses. *Earth and Planetary Science Letters, 530*, 1–13.

International Seabed Authority (ISA). (2022). *Polymetallic sulphides*. International Seabed Authority.

International Seabed Authority (ISA), Khulapova, T. M., Kruglyakova, M. V., Li, H., Ponomareva, I., Weaver, P., Yamaoka, K., Kim, H. J., Nagao, M., Tsukasaki, A., Yuan, Y., Boschen-Rose, R., Baco, A., Du Preez, C., Iguchi, A., Ikeuchi, E., Molodtsova, T., Nishijima, M., Melnik, V., & Xu, X.-W. (2020). *Regional environmental assessment for the Northwest Pacific Ocean*. International Seabed Authority.

International Union for Conservation of Nature (IUCN). (2023, April). *Marine pollution from sunken vessels*. Accessed November 16, 2024, from https://www.iucn.org/resources/issues-brief/marine-pollution-sunken-vessels.

ItvNEWS. (2022, July). Emergency operation after large oil spill off Kent coast averts serious shoreline impact. *ItvNEWS*. https://www.itv.com/news/meridian/2022-07-28/emergency-operation-launched-after-large-oil-spill-threatens-coastline.

Jeffery, B. (2004). World War II shipwrecks in Truk Lagoon: The role of interest groups. *CRM: The Journal of Heritage Stewardship, 1*, 2. https://www.nps.gov/crps/CRMJournal/Summer2004/article2.html

Johnson, A. P., Hannen, W. R., & Zuccari, F. (2013). Vibration control during museum construction projects. *Journal of the American Institute of Conservation, 52*(1), 30–47.

Johnson, D. L., Medlin, D. J., Murphy, L. E., Carr, J. D., & Conlin, D. L. (2011). Corrosion rate trajectories of concreted iron and steel shipwrecks and structures in seawater—The Weins number. *Faculty Publications --- Chemistry Department, 197*, 1–10.

Jones, D. (2023). The Offshore Oil Exploration Industry and Archaeology in the Gulf of Mexico. *Journal of Maritime Archaeology, 18*, 351–369.

Jones, C. A., Reynolds, P., & Pavic, A. (2011). Vibration serviceability of stadia structures subjected to dynamic crowd loads: A literature review. *Journal of Sound and Vibration, 330*, 1531–1566.

Jones, D. O. B., Simon-Lledo, E., Amon, D. J., Bett, B. J., Caulle, C., Clement, L., Connelly, D. P., Dahlgren, T. G., Durden, J. M., Drazen, J. C., Felden, J., Gates, A. R., Georgieva, M. N., Glover, A. G., Gooday, A. J., Hollingsworth, A. L., Horton, T., James, R. H., Jeffreys, R. M., Laguionie-Marchais, C., Leitner, A. B., Lichtschlag, A., Menendez, A., Paterson, G. L. J., Peel, K., Robert, K., Schoening, T., Shulga, N. A., Smith, C. R., Taboada, S., Thurnherr, A. M., Wiklund, H., Young, C. R., & Huvenne, V. A. I. (2021). Environment, ecology, and potential effectiveness of an area protected from deep-sea mining (Clarion Clipperton Zone, abyssal Pacific). *Progress in Oceanography, 197*, 1–19.

Kaikkonen, L., Venesjärvi, R., Nygård, H., & Kuikka, S. (2018). Assessing the impacts of seabed mineral extraction in the deep sea and coastal marine environments: Current methods and recommendations for environmental risk assessment. *Marine Pollution Bulletin, 135*, 1183–1197.

Keith, M. E. (2016). Conclusion. In M. E. Keith (Ed.), *Site formation processes of submerged shipwrecks* (pp. 259–263). University Press of Florida.

Keith, M. E., & Evans, A. M. (2016). Sediment and site formation in the marine environment. In M. E. Keith (Ed.), *Site formation processes of submerged shipwrecks* (pp. 44–69). University Press of Florida.

Kery, S., & Stauffer, J. (2015). Hydrodynamics related to shipwreck Taphonomy. *Proceedings of OCEANS 2015–MTS/IEEE Washington, 1*, 1–21.

Lahaye, N., Gula, J., Thurnherr, A. M., Reverdin, G., Bouret-Aubertot, P., & Roullet, G. (2019). Deep currents in the Rift Valley of the North Mid-Atlantic Ridge. *Frontiers in Marine Science, 6*, 1–17.

Laitinen, T. (2015). *Financial and ecological gains of early intervention in cleaning up wrecks containing oil*. Bachelor's thesis. University of Turku.

Landquist, H., Rosen, L., Lindhe, A., & Hassellöv, I.-M. (2016). VRAKA–A probabilistic risk assessment method for potentially polluting shipwrecks. *Frontiers in Environmental Science, 4*, 1–14.

Landquist, H., Norman, J., Lindhe, A., Norberg, T., Hassellhöv, I.-M., Lindgren, J. F., & Rosen, L. (2017). Expert elicitation for deriving input data for probabilistic risk assessment of shipwrecks. *Marine Pollution Bulletin, 125*(1–2), 399–415.

Lawrence, M., Leather, S., & Burnay, S. (2024). Assessment methodologies for potentially polluting wrecks: The need for a common approach. In M. Brennan (Ed.), *Threats to our ocean heritage: Potentially polluting wrecks* (pp. 143–155). Springer.

Leng, D., Shao, S., Xie, Y., Wang, H., & Liu, G. (2021a). A brief review of recent progress on deep sea mining vehicle. *Ocean Engineering, 228*, 1–17.

Leng, D., Zhu, Z., Xu, K., Li, Y., & Liu, G. (2021b). Vibration control of jacket offshore platform through magnetorheological elastomer (MRE) based isolation system. *Applied Ocean Research*, 1–12.

Levin, L. A., Mengerink, K., Gjerde, K. M., Rowden, A. A., Van Dover, C. L., Clark, M. R., Ramirez-Llodra, E., Currie, B., Smith, C. R., Sato, K. N., Gallo, N., Sweetman, A. K., Lily, H., Armstrong, C. W., & Brider, J. (2016). Defining "serious harm" to the marine environment in the context of deep seabed mining. *Marine Policy, 74*, 245–259.

Lin, T.-H., Chen, C., Watanabe, H. K., Kawagucci, S., Yamamoto, H., & Akamatsu, T. (2019). Using soundscapes to assess deep-sea benthic ecosystems. *Trends in Ecology & Evolution, 34*(12), 1066–1069.

Lingren, J. F., Åberg, F., & Johansson, J. R. (2021). Polluting shipwrecks in Swedish waters: Investigations, risk assessment methodology and oil removal operations. *2021 International Oil Spill Conference Proceedings, 1*, 1–20.

Liu, Z., Liu, K., Chen, X., Ma, Z., Lv, R., Wei, C., & Ma, K. (2023). Deep-sea rock mechanics and mining technology: State of the art and perspectives. *International Journal of Mining Science and Technology, 33*(9), 1083–1115.

Loureiro, M. L., Ribas, A., López, E., & Ojea, E. (2006). Estimated costs and admissible claims linked to the Prestige oil spill. *Ecological Economics, 59*(1), 48–63.

Lu, C., Yang, J., Leira, B. J., Chen, Q., & Wang, S. (2023). Three-dimensional path planning of deep-sea mining vehicle based on improved particle swarm optimization. *Journal of Marine Science and Engineering, 11*(9), 1–26.

Luciano, G., Letardi, P., Traverso, P., & Belsanti, L. (2013). Corrosion behavior of Al, Cu, and Fe alloys in deep sea environment. *La metallurgia italiana, 1*, 21–29.

Luckenbach Trustee Council, California Department of Fish and Game, National Oceanic and Atmospheric Administration, United States Fish and Wildlife Service, & National Park Service (LTC). (2006). *S.S. Jacob Luckenbach and associated mystery oil spills final damage assessment and restoration plan/environmental assessment*. California Department of Fish and Game.

Lyle, M. (2015). Deep-sea sediments. In J. Harff, M. Meschede, S. Petersen, & J. Thiede (Eds.), *Encyclopedia of marine geosciences* (pp. 1–20). Springer.

Ma, M., Liu, W., Qian, C., Deng, G., & Li, Y. (2016). Study of the train-induced vibration impact on a historic Bell Tower above two spatially overlapping metro lines. *Soil Dynamics and Earthquake Engineering, 81*, 58–74.

MacLeod, I. D. (2002). In situ corrosion measurements and management of shipwreck sites. In C. V. Ruppe & J. F. Barstad (Eds.), *International handbook of underwater archaeology* (pp. 697–714). Springer.

MacLeod, I. D. (2019). Corrosion and conservation management of the submarine HMAS AE2 (1915) in the Sea of Marmara, Turkey. *Heritage, 2*, 868–883.

MacLeod, I. D., Selman, A., & Selman, C. (2017). Assessing the impact of typhoons on historic iron shipwrecks in Chuuk Lagoon through changes in the corrosion microenvironment. *Conservation and Management of Archaeological Sites, 19*(4), 269–287.

MacLeod, I. D., & Steyne, H. (2011). Managing a monitor—The case of HMVS Cerberus in Port Phillip Bay: Integration of corrosion measurements with site management strategies. *Conservation and Management of Archaeological Sites, 13*(4), 334–361.

MacLeod, I. D., Richards, V., & Beger, M. (2011). The effects of human and biological interactions on the corrosion of WWII iron shipwrecks in Chuuk Lagoon. *Proceedings from the 18th International Corrosion Congress, 1*, 1–12.

Manglis, A., Fourkiotou, A., & Papadopoulou, D. (2021). A roadmap for the sustainable valorization of accessible underwater cultural heritage sites. *Heritage, 4*, 4700–4715.

Martin, C., Weilgart, L., Amon, D. J., & Müller, J. (2021). *Deep-Sea mining: A noisy affair. Overview and recommendations*. OceanCare. https://www.oceancare.org/wp-content/uploads/2021/11/OceanCare_a-noisy-affair_pdf.pdf.

Marx, D. E., & Delgado, J. P. (2013). *World War II shipwrecks along the East Coast and Gulf of Mexico: National register of historic places multiple property nomination*. NOAA's Office of National Marine Sanctuaries. Accessed October 26, 2024, from https://nmsmonitor.blob.core.windows.net/monitor-prod/media/docs/20130426-national-register-of-historic-places-multiple-property.pdf

Masetti, G. (2012). *A geo-database for potentially polluting marine sites and associated risk index*. Master's Degree Thesis. University of New Hampshire.

Mauro, M. (2020). *Mining extraction in the ocean depths: A baseline to understand and reduce acoustic impact on biodiversity*. Doctoral Thesis. Universitat Politècnica de València.

McCarthy, M. (2000). *Iron and steamship archaeology: Success and failure on the SS Xantho*. Kluwer Academic/Plenum Publishers.

Mejjad, N., & Rovere, M. (2021). Understanding the impacts of blue economy growth on deep-sea ecosystem services. *Sustainability, 13*, 1–26.

Melchers, R. E. (2003). Probabilistic models for corrosion in structural reliability assessment—Part 2: Models based on mechanics. *Journal of Offshore Mechanics and Arctic Engineering, 125*, 272–280.

Michel, J., Gilbert, T., Waldron, J., Blocksidge, C. T., Etkin, D. S., & Urban, R. (2005). Potentially polluting wrecks in marine waters: An issue paper prepared for the 2005 International Oil Spill Conference. *Annals of the 2005 International Spill Conference*, 1–77.

Mining.com. (2024). *Chinese deep-sea mining vehicle reaches 4,000m depth in trial run*. Mining.com. Accessed August 29, 2024, from https://www.mining.com/chinese-deep-sea-vehicle-reaches-4000m-depth-in-trial-run/.

Moller, T. H., Parker, H. D., & Nichols, J. A. (1987). Comparative costs of oil spill cleanup techniques. *Proceedings of the International 1987 Oil Spill Conference, 1*, 123–127.

Monti, G., Fumagalli, F., Marano, G. C., Quaranta, G., Rea, R., & Nazzaro, B. (2013). Effects of ambient vibrations on heritage buildings: Overview and wireless dynamic monitoring application. In C. Baggio, P. Clemente, Y. Nakamura, L. Orlando, A. Rovelli, & G. Valente (Eds.), *Dynamic Interaction of Soil and Structure (Diss_13). Proceedings of the 3rd International Workshop: Dynamic Interaction between Soil, Monuments, and Built Environment* (pp. 217–229). L'Aquila University.

Moore, C. (2021). *The practicalities of managing and mapping potentially polluting shipwrecks in the UK: Legal, social and ethical considerations*. PhD thesis. University of Southampton.

Moore, J. D., III. (2015). Long-term corrosion processes of iron and steel shipwrecks in the marine environment: A review of current knowledge. *Journal of Maritime Archaeology, 10*, 191–204.

Motion & Flow Control Products, Inc. (n.d.). *Galvanic Corrosion Scale | Corrosion of Base Metals in Contact*. Accessed October 19, 2024, from https://www.mfcp.com/technical-info/galvanic-corrosion.

Mugge, R. L., Brock, M. L., Salerno, J. L., Damour, M., Church, R. A., Lee, J. S., & Hamdan, L. J. (2019). Deep-sea biofilms, historic shipwreck preservation and the deepwater horizon spill. *Frontiers of Marine Science, 13*(6), 1–17.

Muñoz-Royo, C., Peacock, T., Alford, M. H., Smith, J. A., Le Boyer, A., Kulkarni, C. S., Lermusiaux, P. F. J., Haley, P. J., Jr., Mirabito, C., Wng, D., Adams, E. E., Ouillon, R., Breugem, A., Decrop, B., Lanckriet, T., Supekar, R. B., Rzeznik, A. J., Gartman, A., & Ju, S.-J. (2021). Extent of impact of deep-sea nodule mining midwater plumes is influenced by sediment loading, turbulence and thresholds. *Communications: Earth and Environment.*, 1–16.

Napang, M., Patittingi, F., Hambali, R., & Hendrapati, M. (2019). Safety zones around offshore installations. *International Journal of Global Community, 2*(2), 139–150.

National Oceanic and Atmospheric Administration (NOAA). (2013a). *Risk assessment for potentially polluting wrecks in U.S. waters*. Accessed October 26, 2024, from https://nmssanctuaries.blob.core.windows.net/sanctuariesprod/media/archive/protect/ppw/pdfs/2013_potentiallypollutingwrecks.pdf.

National Oceanic and Atmospheric Administration National Marine Sanctuaries (NOAA). (2013b). *Potentially polluting wrecks in U.S. waters*. National Marine Sanctuaries. Accessed November 17, 2024, from https://sanctuaries.noaa.gov/protect/ppw/wrecks.html.

National Oceanic and Atmospheric Administration (NOAA) Fisheries. (2018). *Fishing gear: Dredges*. NOAA: Department of Commerce. Accessed August 29, 2024, from https://www.fisheries.noaa.gov/national/bycatch/fishing-gear-dredges#:~:text=Dredge%20construction%2C%20tow%20time%2C%20and,15%20minute%20tows.

National Research Council (US) Committee on Oil in the Sea: Inputs, Fates, and Effects. (2003). *Oil in the sea III: Inputs, fates, and effects*. National Academies Press.

Ngundu, K., Beylich, B. A., Staalstrøm, A., Øxnevad, S., Berge, J. A., Braaten, H. F. V., Schaanning, M., & Bergstrøm, R. (2017). Petroleum oil and mercury pollution from shipwrecks in Norwegian coastal waters. *Science of the Total Environment, 593–594*, 624–633.

Noble, M., & Mullineaux, L. S. (1989). Internal tidal currents over the summit of Cross Seamount. *Deep-Sea Research, 36*(12), 1791–1802.
North, N. A. (1976). Formation of coral concretions on marine iron. *The International Journal of Nautical Archaeology and Underwater Exploration, 5*(3), 253–258.
North, N. A. (1984). The role of galvanic couples in the corrosion of shipwreck metals. *International Journal of Nautical Archaeology, 13*(2), 97–188.
North, N. A., & MacLeod, I. D. (1987). Corrosion of metals. In E. C. Pearson (Ed.), *Corrosion of Marine Archaeological Objects* (pp. 68–98). Butterworths.
Northey, H. (2023). *The world's moving closer to deep-sea mining. There are no rules.* E&E News by Politico. Accessed October 20, 2024, from https://www.eenews.net/articles/the-worlds-moving-closer-to-deep-sea-mining-there-are-no-rules/.
Office of Resources and Environmental Monitoring. (2006). *Mining cobalt-rich ferromanganese crusts and polymetallic sulphides deposits: Technological and economic considerations.* International Seabed Authority.
Papageorgiou, M. (2019). Stakes and challenges for underwater cultural heritage in the era of blue growth and the role of spatial planning: Implications and prospects in Greece. *Heritage, 2*, 1060–1069.
Paxton, A. B., McGonigle, C., Damour, M., Holly, G., Caporaso, A., Campbell, P. B., Meyer-Kaiser, K. S., Hamdan, L. J., Mires, C. H., & Taylor, J. C. (2024). Shipwreck ecology: Understanding the function and processes from microbes to megafauna. *BioScience, 74*, 12–24.
Petersen, S., Krätschell, A., Augustin, N., Jamieson, J., Hein, J. R., & Hannington, M. D. (2016). News from the seabed–Geological characteristics and resource potential of deep-sea mineral resources. *Marine Policy, 70*, 175–187.
Pfeiffer, F. (2012). Changes in properties of explosives due to prolonged seawater exposure. *Marine Technology Society Journal, 46*(1), 102–110.
Pickens, C., Lily, H., Harrould-Kolieb, E., Blanchard, C., & Chakraborty, A. (2024). From what-if to what-now: Status of the deep-sea mining regulations and underlying drivers for outstanding issues. *Marine Policy, 169*, 1–19.
Pollard, R., & Read, J. (2017). Circulation, stratification and seamounts in the Southwest Indian Ocean. *Deep Sea Research Part II: Topical Studies in Oceanography, 136*, 36–43.
Przeslawski, R., & Christenhusz, M. J. M. (2022). Editorial: Deep-sea discoveries. *Zoological Journal of the Linnean Society, 194*, 1037–1043.
Quinn, R. (2006). The role of scour in shipwreck site formation processes and the preservation of wreck-associated scour signatures in the sedimentary record – evidence from seabed and sub-surface data. *Journal of Archaeological Science, 33*, 1419–1432.
Radziejewska, T., Węsławski, J. M., Tomczak, M., Błachowiak-Samołyk, K., Błażewicz, M., Damrat, M., Jankowska, K., Jóźwiak, P., Kotrys, B., Kozłowska-Roman, A., Linkowski, T. B., Maszloch, E., Pabis, K., Piskozub, J., Sagan, S., Smajdor, L., Stempniewicz, L., Sydor, P., Szczuciński, W., Szyduk, A., Walczowski, W., Włodarska-Kowalczuk, M., & Żuk, T. (2020). *Environment of the Mid-Atlantic Ridge area under the contract for polymetallic sulphide exploration between the International Seabed Authority and the Ministry of Environment of the Republic of Poland.* Ministry of Environment of the Republic of Poland.
Ramirez-Llodra, E., Brandt, A., Danovaro, R., De Mol, B., Escobar, E., German, C. R., Levin, L. A., Martinez Arbizu, P., Menot, L., Buhl-Mortenson, P., Narayanaswamy, B. E., Smith, C. R., Tittensor, D. P., Tyler, P. A., Vanreusel, A., & Vecchione, M. (2010). Deep, diverse and definitely different: Unique attributes of the world's largest ecosystem. *Biogeosciences, 7*, 2851–2899.
Research Planning, Inc, Tidewater Atlantic Research, Inc, & W.F. Baird & Associates Ltd. (2004). *Archaeological damage from offshore dredging: Recommendations for pre-operational surveys and mitigation during dredging to avoid adverse impacts.* Research Planning, Inc. https://www.boem.gov/sites/default/files/mm-research/2022-03/2004-005.pdf
Ridwan, N. N. H. (2019). Vulnerability of shipwreck sites in Indonesian waters. *Current Science, 117*(10), 1623–1628.

Ruggieri, S., Bruno, G., Attolico, A., & Uva, G. (2024). Assessing the dredging vibrational effects on surrounding structures: The case of port nourishment in Bari. *Journal of Building Engineering, 96*, 1–19.

Saadoun, I. M. K. (2015). Impacts of oil spills on marine life. In M. L. Larramendy & S. Soloneski (Eds.), *Emerging pollutants in the environment: Current and further implications* (pp. 77–105). Intech.

Salerno, J. L., Little, B., Lee, J., & Hamdan, L. J. (2018). Exposure to crude oil and chemical dispersant may impact marine microbial biofilm composition ABD steel corrosion. *Frontiers in Marine Science, 5*, 1–14.

Schjølberg, I., & Utne, I. B. (2015). Towards autonomy in ROV operations. *IFAC-PapersOnLine, 48*(2), 183–188.

Secretariat for the International Seabed Authority Council. (2006). *Exploration and mine site model applied to block selection for cobalt-rich ferromanganese crusts and polymetallic sulphides. Part 1: Cobalt-rich ferromanganese crusts. ISBA/12/C/3: Part I.* International Seabed Authority.

Sharma, R. (2011). Deep-sea mining: Economic, technical, technological, and environmental considerations for sustainable development. *Marine Technology Society Journal, 45*(5), 28–41.

Sharma, R. (2020). Potential impacts of deep-sea mining on ecosystems. *Oxford Research Encyclopedia of Environmental Science*, 1–37.

Sharma, R., Nagender Nath, B., Ingole, B., Murty, V. S. N., Khadge, N. H., Mascarenhas-Pereira, M. B., Babu, M. T., Murali, M., Sankar, S. J., Patil, J., Gaikwad, V., Atmanand, M. A., Rmadass, G. A., Gopkumar, K., Kumar, V., & Singh, T. R. P. (2020). *Environmental impact statement: Environmental conditions and likely impact in the area selected for nodule collection trials at the Indian PMN site in the Central Indian Ocean Basin.* Ministry of Earth Sciences of the Government of India.

Smyth, A. W., Brewick, P., Greenbaum, R., Chatzis, M., Serotta, A., & Stünkel, I. (2016). Vibration mitigation and monitoring: A case study of construction in a museum. *Journal of the American Institute for Conservation, 55*(1), 32–55.

South Pacific Regional Environment Programme (SPREP). (2002). *A regional strategy to address marine pollution from World War II wrecks*. SPREP.

Souza-Neto, P. W. M., da Silveira, I. C. A., Rocha, C. B., Lazaneo, C. Z., & Calil, P. H. R. (2024). The Rio Grande Rise circulation: Dynamics of an internal tide conversion hotspot in the Southwestern Atlantic. *Progress in Oceanography, 224*, 1–13.

Sparenberg, O. (2019). A historical perspective on deep-sea mining for manganese nodules, 1965–2019. *The Extractive Industries and Society, 6*, 842–854.

Spearman, J., Taylor, J., Crossouard, N., Cooper, A., Turnbull, M., Manning, A., Lee, M., & Murton, B. (2020). Measurement and modelling of deep sea sediment plumes and implications for deep sea mining. *NatureResearch Scientific Reports, 10*, 1–14.

Step Change in Safety. (2017). *Marine operations: 500m safety zone*. Accessed October 20, 2024, from https://www.marinesafetyforum.org/wp-content/uploads/2018/08/Marine-Operations-500m-zone-guidance.pdf.

Stokesbury, K., Baker, E., Harris, B., & Rheault, R. (2011). Environmental impacts related to mechanical harvest of cultured shellfish. In S. Shumway (Ed.), *Shellfish aquaculture and the environment* (pp. 319–338). Wiley-Blackwell.

The Life-Boat. (1947). Tanker on fire. *The Journal of the Royal National Life-boat Institution, 32*(343), 68–69. https://rnliarchive.blob.core.windows.net/media/1344/0343.pdf#page=36

Tripathy, G. R., & Shirke, R. R. (2015). Underwater drilling and blasting for hard rock dredging in Indian ports – A case study. *Aquatic Procedia, 4*, 248–255.

Van Dover, C. L. (2014). Impacts of anthropogenic disturbances at deep-sea hydrothermal vent ecosystems: A review. *Marine Environmental Research, 102*, 59–72.

Ventikos, N., Koimtzoglou, A., & Louzis, K. (2014). Shipwreck: A crisis with challenging solutions. *Journal of Risk Analysis and Crisis Response, 4*(3), 160–174.

Victorovna, M. L. (2018). Vibration strength of pipelines. In T. Nakamura (Ed.), *System of system failures*. InTechOpen. Accessed November 7, 2024, from https://www.intechopen.com/chapters/58981

Vineesh, T. C., Nagender Nath, B., Banerjee, R., Jaisankar, S., & Lekshmi, V. (2009). Manganese nodule morphology as indicators for oceanic processes in the Central Indian Basin. *International Geology Review, 51*(1), 27–44.

Washburn, T. W., Turner, P. J., Durden, J. M., Jones, D. O. B., Weaver, P., & Van Dover, C. L. (2019). Ecological risk assessment for deep-sea mining. *Ocean and Coastal Management, 176*, 24–39.

Washburn, T. W., Jones, D. O. B., Wei, C.-L., & Smith, C. R. (2021). Environmental heterogeneity throughout the Clarion-Clipperton zone and the potential representativity of the APEI network. *Frontiers in Marine Science, 8*, 1–17.

Webb, P. (2023). *Introduction to oceanography*. Roger Williams University.

Wei, W., Watts, S., Seddon, T., & Crombie, D. (2018). Protecting museum collections from vibrations due to construction: Vibration statistics, limits, flexibility and cooperation. *Studies in Conservation, 63*(1), 293–300.

Weilgart, L. (2017). Din of the deep: Noise in the ocean and its impacts on Cetaceans. In A. Buttersworth (Ed.), *Marine mammal welfare* (pp. 111–124). Springer.

Wenhai, L., Cusack, C., Baker, M., Tao, W., Mingbao, C., Paige, K., Xiaofan, Z., Levin, L., Escobar, E., Amon, D., Yue, Y., Reitz, A., Sepp Neves, A. A., O'Rourke, E., Mannarini, G., Pearlman, J., Tinker, J., Horsburgh, K. J., Lehodey, P., Pouliquen, S., Dale, T., Peng, Z., & Yufeng, Y. (2019). Successful blue economy examples with an emphasis on international perspectives. *Frontiers in Marine Science, 6*, 1–14.

Wilkinson, D., & Coroneos, C. (2017). *Underwater cultural heritage and seabed mining in the Northern Territory with applicability to other marine industries: Strategy for management*. Northern Territory Government Department of Tourism and Culture: Heritage Branch.

Williams, R., Erbe, C., Dunca, A., Nielsen, K., Washburn, T., & Smith, C. (2022). Noise from deep-sea mining may span vast ocean areas. *Science, 377*, 157–158.

Wright, J. (2016). Maritime archaeology and climate change: An invitation. *Journal of Maritime Archaeology, 11*, 255–270.

Xie, Y., Gao, C., Wang, P., Zhou, L., Zhang, C., & Qu, X. (2023). Research on vibration fatigue damage locations of offshore oil and gas pipelines based on the GA-improved BP neural network. *Shock and Vibration*, 1–18.

Xu, W., Zhao, J., Yang, W., Yu, D., Qiu, L., Chen, Z., & Qiu, Y. (2023). The effects of construction vibration on stone cultural relics: A case study from the Qingdao Museum. *Heritage Science, 11*(250), 1–18.

Yanagimoto, D., Kawabe, M., & Fujio, S. (2010). Direct velocity measurements of deep circulation southwest of the Shatsky Rise in the western North Pacific. *Deep Sea Research Part I: Oceanographic Research Papers, 57*(3), 328–337.

Zhang, Q., Chen, X., Luan, L., Sha, F., & Liu, X. (2024). Technology and equipment of deep-sea mining: State of the art and perspectives. *Earth Energy Science*, 1–20. (In Press, Corrected Proof).

Zhang, T., Wang, W., Li, X., & Wang, B. (2023). Vibration mitigation in offshore wind turbine under combined wind-wave-earthquake loads using the tuned mass damper inerter. *Renewable Energy, 216*, 1–17.

Zhu, L., Meng, J., Chen, L., & Hu, X. (2023). Experimental investigation and in-situ testing of traffic-induced vibrations on the adjacent ruins of an ancient cultural sites. *Applied Sciences, 13*(24), 1–17.

Zigic, S. (2004). Ocean modelling–systems available for the environment and scientific coordinator. *Proceedings from the 2004 environment and scientific coordinators workshop, 1*, 49–71.

Zini, G., Betti, M., & Bartoli, G. (2021). Experimental analysis of the traffic-induced-vibration on an ancient lodge. *Structural Control Health Monitoring, 29*, 1–19.

Map Sources

Global Fishing Watch. (2021). *Datasets and code: Fishing effort*. Global Fishing Watch. Accessed December 1, 2024, from https://globalfishingwatch.org/data-download/datasets/public-fishing-effort.

Heersink, P. (2022). *Sunken ships of the Second World War*. Esri. Accessed July 29, 2024, from https://www.arcgis.com/apps/dashboards/fe88b5e18c6443c7afaf6e32f8432687.

International Seabed Authority (ISA). (2024). *Maps*. International Seabed Authority. Accessed November 17, 2024, from https://www.isa.org.jm/exploration-contracts/maps/.

Kroodsma, D. A., Mayorga, J., Hochberg, T., Miller, N. A., Boerder, K., Ferretti, F., Wilson, A., Bergman, B., White, T. D., Block, B. A., Woods, P., Sullivan, B., Costello, C., & Worm, B. (2018). Tracking the global footprint of fisheries. *Science, 359*(6378), 904–908.

Miller, K. A., Thompson, K. F., Johnston, P., & Santillo, D. (2018). An overview of seabed mining including the current state of development, environmental impacts, and knowledge gaps. *Frontiers of Marine Science, 4*, 1–24.

UNEP-WCMC, & IUCN (2024). *Protected planet: The world database on protected areas (WDPA)*. Accessed November 17, 2024, from www.protectedplanet.net.

Image Sources

Dillon, A. (2020). Mining-generated sediment plumes and noise have a variety of possible effects on pelagic taxa. (Organisms and plume impacts are not to scale.). [Digital]. In Drazen, J. C., Smith, C. R., Gjerde, K. M., Haddock, S. H. D., Carter, G. S., Choy, C. A., Clark, M. R., Dutrieux, P., Goetze, E., Hauton, C., Hatta, M., Koslow, J. A., Leitner, A. B., Pacini, A., Perelman, J. N., Peacock, T., Sutton, T. T., Watling, L., & Yamamamoto, H. (2020). Midwater ecosystems must be considered when evaluating environmental risks of deep-sea mining. *Proceedings of the National Academy of Sciences, 117*(30), 17455–17460.

Maldjian, S. (2020). *Rocks*. Noun Project. Accessed November 16, 2024, from https://thenounproject.com/icon/rocks-3575771/.

Oki Orlando, M. (2022). *Seamount*. Noun Project. Accessed November 16, 2024, from https://thenounproject.com/icon/seamount-5399400/.

Wu, K. (2016). *Chimney*. Noun Project. Accessed November 16, 2024, from https://thenounproject.com/icon/chimney-763381/.

Chapter 7
Deep Sea Mining and New Zealand's Underwater Cultural Heritage

Kurt Bennett and Melanie Russell

Abstract As interest and opportunity in deep-sea mineral extraction grows, the increased activity poses a new threat to the underwater cultural heritage (UCH) sites in New Zealand's deep-sea environment. Currently this environment is largely unexplored in terms of professional maritime archaeological investigation and recording, however, there are some significant shipwrecks such as the SS *Ventnor,* RMS *Niagara*, and SS *Port Kembla*, that have been explored by technical divers. Sites like these are both fragile and historically valuable and as such, current protection measures and potential adverse effects arising from seabed mining need to be assessed in order to understand how to offer some protection to the valuable information that lies on and below the sea floor. This chapter examines the potential risks to UCH posed by offshore resource development and reviews New Zealand's current legislative framework, including the *Resource Management Act 1991*, the *Heritage New Zealand Pouhere Taonga Act 2014*, and the *Exclusive Economic Zone and Continental Shelf (Environmental Effects) Act 2012*. It highlights the regulatory gaps in the protection of UCH beyond the 12 nautical mile limit and underscores the need for urgent policy reform. With maritime archaeology still emerging as a professional discipline in New Zealand, the chapter calls for increased investment in site surveys, regulatory expertise, and public education. It advocates for a precautionary and proactive approach to offshore development that prioritises the identification, documentation, and protection of UCH as a critical component of New Zealand's maritime heritage.

K. Bennett (✉)
Past Heritage, Auckland, New Zealand
e-mail: kurt.bennett@pastheritage.co.nz

M. Russell
Heritage New Zealand Pouhere Taonga, Christchurch, New Zealand
e-mail: mrussell@heritage.org.nz

C. Jarvis (ed.), *Threats to Our Ocean Heritage: Deep Sea Mining*,
SpringerBriefs in Archaeology, https://doi.org/10.1007/978-3-031-98238-5_7

Keywords Underwater Cultural heritage · Seabed mining · Maritime archaeology · New Zealand · Shipwrecks · Environmental legislation · Exclusive Economic Zone · Heritage protection

7.1 Introduction

New Zealand, located in the South Pacific, is renowned for its natural landscapes and important maritime history. Its territorial waters extend 12 nautical miles (nm) from the coast, while its Exclusive Economic Zone (EEZ) stretches 200 nm, making it one of the largest maritime zones to manage in the world (Fig. 7.1). The country's sea boundaries encompass a diverse range of marine environments, from shallow coastal areas to the deep sea, which is home to unique ecosystems and significant underwater cultural heritage (UCH). The UNESCO 2001 Convention on the Protection of Underwater Cultural Heritage defines UCH as all traces of human existence that lie or have lain underwater and are of a cultural or historical nature. New Zealand, however, has not ratified this convention.

The deep-sea environment around New Zealand is characterised by its cold, dark waters and high pressure, creating a habitat for a variety of marine life, including unique species. In these volumes of water also lies important UCH which has not been relocated or accessed in the past. Meanwhile, the discipline of maritime archaeology in New Zealand, while historically underdeveloped, has seen notable improvements over the last decade (Carter & Bennett, 2022). Increased research and exploration efforts have begun to uncover and document UCH, although much work remains to be done to fully understand and protect sites, particularly in deeper waters.

Seabed mining is becoming increasingly prominent as new areas of seafloor and particularly the deep sea is explored for precious metals and minerals. The International Union for Conservation of Nature (2022) describes seabed mining as the process of retrieving mineral deposits from the deep seabed (below 200 m). However, New Zealand's legislative framework relating to seabed mining is not depth related, rather it regulates the activities depending on distances from the shoreline. As this industry develops in New Zealand's waters, it poses both opportunities and challenges for the preservation of UCH.

New Zealand's legislative framework for protecting UCH is multi-layered, involving several national and international laws and regulations (Davies & Myburgh, 2006; see Bennett & Parslow, 2025). This framework provides for the protection and management of heritage sites, ensuring their preservation for future generations. The current state of deep underwater cultural heritage in New Zealand is one of both preservation and vulnerability. While many shallower sites continue to be documented and protected, others remain at risk from natural and human-induced threats.

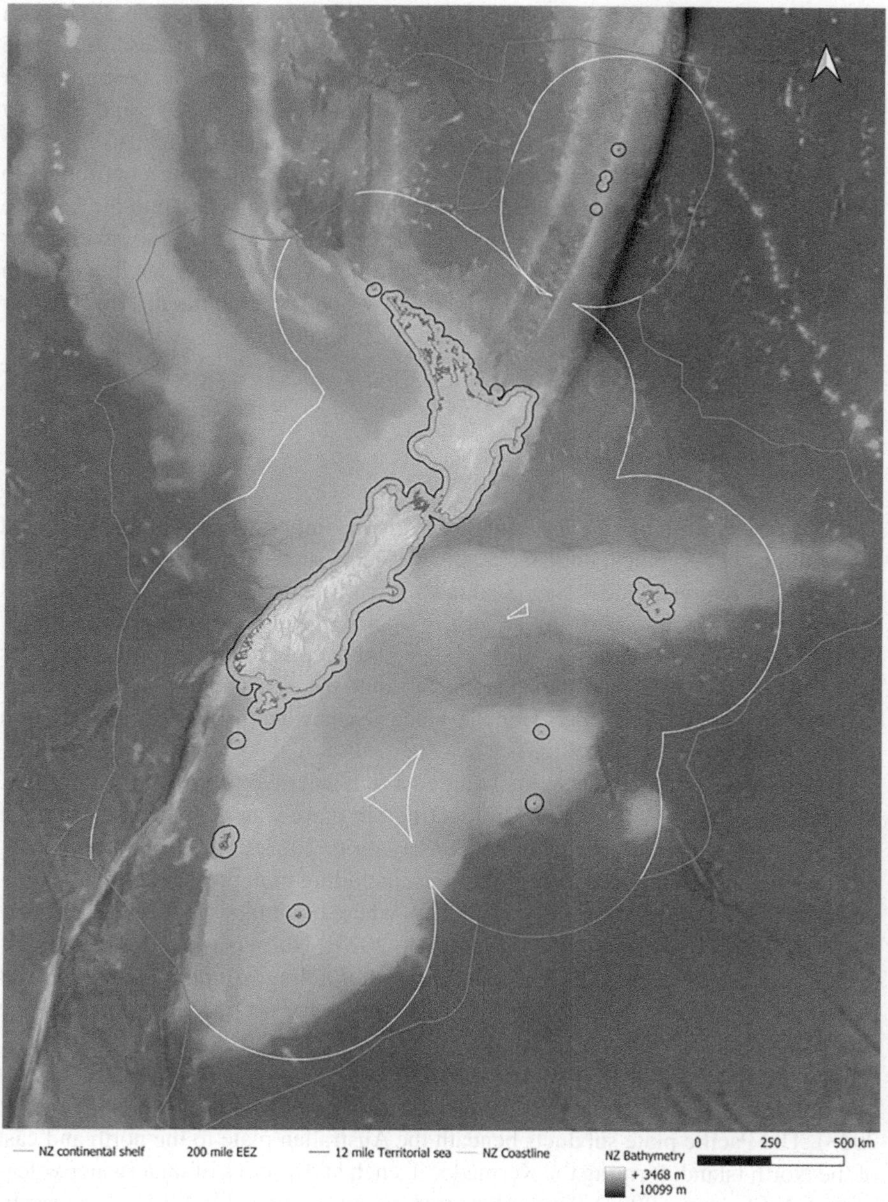

Fig. 7.1 Map of New Zealand's territorial boundaries, EEZ and continental shelf. (Source: Authors' own)

Maritime heritage holds immense importance for the island nation with a rich seafaring tradition. The waters surrounding New Zealand are home to numerous significant sites that reflect its maritime history and cultural connections. These sites not only provide historical insights but also contribute to the nation's identity and heritage, making their preservation crucial.

This chapter explores the potential threats and opportunities associated with deep sea mining and its impact on UCH in New Zealand. By briefly describing case studies such as SS *Ventnor* (sank in 1902 while carrying the human remains of Chinese miners), RMS *Niagara* (sunk by a German mine in 1940), and SS *Port Kembla* (sunk by a mine in 1917), it highlights the importance of safeguarding these underwater treasures and the need to assess others. The following contributes to New Zealand's offshore future development by providing insights into the balance between preservation and progress and offers a comparable case study with other international jurisdictions. Since New Zealand deep-water UCH has never been specifically described in the context of legislative and potential threats, this chapter establishes an integral foundation for future considerations.

7.2 Deep-Sea Environment

The deep sea surrounding New Zealand is a vast and important environment, defined by its unique geographical boundaries, diverse ecosystems, and dynamic geological activity. This region includes New Zealand's Territorial Sea and EEZ, which extends from 12 nm to 200 nm, covering an area of approximately 4.2 million square kilometres (Department of Conservation, 2023). The deep sea here encompasses parts of the Tasman Sea to the west, the Pacific Ocean to the east, and notable underwater features such as the Kermadec Trench, which reaches depths exceeding 10,000 m (Linley et al., 2017; NIWA, 2023).

The deep sea of New Zealand is home to a rich and diverse array of marine life. More than 17,000 species have been identified in these waters, many of which are unique to the region (Department of Conservation, 2023). The deep-sea environment is characterised by extreme conditions, including high pressure, low temperatures, and complete darkness beyond 1000 m, where bioluminescence is the primary source of light (Department of Conservation, 2023). Nutrients in these depths primarily come from marine snow—organic material falling from the upper layers of the ocean—and from hydrothermal vents and methane seeps that support unique ecosystems (Department of Conservation, 2023).

New Zealand's location on the boundary between the Pacific and Australian tectonic plates makes it a hotspot for volcanic and tectonic activity (GNS Science, 2023). The Pacific plate subducts beneath the Australian plate to the north and east of the North Island, forming the Kermadec Trench and a series of underwater volcanoes (GNS Science, 2023). This subduction zone is responsible for frequent earthquakes and volcanic eruptions, which continuously reshape the seafloor. The South Island features the Alpine Fault, where the plates slide past each other horizontally (GNS Science, 2023). These geological processes contribute to the dynamic nature of New Zealand's deep sea, with new landforms constantly being created and old ones being destroyed.

The deep sea supports a variety of life forms adapted to its harsh conditions. In the mesopelagic zone (200–1000 m), species such as lanternfish and bristle mouths are common, while the bathypelagic zone (1000–4000 m) hosts creatures like giant squid and deep-sea anglerfish (Department of Conservation, 2023). The abyssopelagic zone (4000–6000 m) and hadal zone (below 6000 m) are home to even more specialised organisms, including amphipods and certain species of sea cucumbers (Department of Conservation, 2023). These organisms rely on the sparse nutrients available, often deriving energy from chemosynthesis at hydrothermal vents (Department of Conservation, 2023).

The chemical composition of the deep sea around New Zealand is influenced by various factors, including the input of organic material from the surface and the presence of hydrothermal vents (Te Ara, 2023a, 2023b). Sediments in these depths are primarily composed of fine particles from land, biogenic material from dead plankton, and minerals precipitated from seawater (Te Ara, 2023a, 2023b). The deep sea also contains significant amounts of dissolved gases, such as carbon dioxide and methane, which play a role in global biogeochemical cycles (Te Ara, 2023a, 2023b).

The seafloor morphology around New Zealand is a mosaic of geological features, including continental shelves, slopes, rugged seamounts, abyssal plains, and deep trenches (Te Ara, 2023a, 2023b). The continental shelf is relatively shallow and rich in marine life and minerals. In contrast, the continental slope, which descends steeply to the deep ocean basins, is dissected by submarine canyons that transport sediments from the shelf to the abyssal plains (Te Ara, 2023a, 2023b). The Kermadec Trench, the second deepest ocean trench in the world, is a significant feature of the seafloor, hosting unique ecosystems adapted to its extreme conditions (Te Ara, 2023a, 2023b). Sediments on the seafloor vary widely, from coarse sands and gravels near the shore to fine clays and biogenic oozes in the deep ocean (Te Ara, 2023a, 2023b). These sediments are transported by ocean currents, turbidity currents, and submarine landslides, creating a dynamic and constantly changing seafloor landscape.

New Zealand's deep sea is rich in mineral resources, including significant deposits of metals and sands targeted for deep sea mining. The South Taranaki Bight, for example, contains large quantities of iron sands, which are mined for their high iron content (Mining Digital, 2023). The Chatham Rise is known for its phosphate nodules, which are valuable for agricultural fertilisers (Te Ara, 2023a, 2023b). Additionally, the margins of the Campbell Plateau and the Kermadec volcanic arc contain manganese nodules and crusts rich in manganese, iron, copper, nickel, and cobalt (Te Ara, 2023a, 2023b). These resources are of great economic interest, but their extraction poses environmental challenges and requires careful management to minimise impacts on the deep-sea ecosystem.

New Zealand's deep sea is an important and complex environment, shaped by its unique geological setting and home to diverse marine life. Its rich mineral resources offer significant economic potential but also highlight the need for sustainable management practices to protect this fragile and largely unexplored frontier, including UCH found at its great depths.

7.3 Underwater Cultural Heritage in New Zealand

New Zealand boasts a variety of UCH sites. These sites are found in estuaries, rivers, lakes, and swamps, reflecting the diverse history of both Māori and Pākehā (non-Māori of European origin). The Polynesian peoples, who later became known as Māori, settled across the country, utilising its natural resources and developed a vibrant culture that blended elements from their Pacific homeland with their new environment (Smith, 2019). Archaeological remains of Māori culture have been discovered underwater and in intertidal zones, including canoe (waka) landings, eel weirs, fish traps, inundated settlements, and waka (Carter & Dodd, 2015; Johns et al., 2014).

The UCH of Pākehā dates back to the period of maritime exploration and colonisation from 1769 onwards. This includes approximately 2000 documented shipwrecks, of which only about approximately 10% have been relocated, as well as military sites, wharves, navigation markers, ballast dumps, slipways, and debris (Dodd, 2003, p. 151). Pākehā settlers adapted their seafaring practices to the new environment, creating unique forms of maritime material culture (Carter, 2019). The expansion of the British and French colonial Empires and global trade brought ships from around the world to New Zealand, enriching its archaeological record with diverse cultural resources. These historic resources offer significant opportunities for maritime and underwater archaeology, and they also provide comparative material for terrestrial archaeology.

The introduction of SCUBA diving in New Zealand in the mid-1950s led to the discovery and looting of many UCH sites. This trend dominated diving history in New Zealand until recent years, when a preservation ethos began to take hold. The Maritime Archaeological Association of New Zealand (MAANZ), established in 1989, has played a role in promoting the protection of these sites (Churchill, 1991). MAANZ volunteers have surveyed several notable underwater and maritime heritage sites, including the Mahanga Bay Wharf, the Inconstant project, the shipwreck of *Hydrabad*, and the Armed Constabulary whaleboat in Lake Waikaremoana (Carter & Dodd, 2015). Another advocational group, the Underwater Heritage Group, also conducts maritime heritage projects. A notable project was the relocation of SS *Ventnor* and is described later in this chapter.

Since 2002, the Australasian Institute for Maritime Archaeology (AIMA) has partnered with the Nautical Archaeology Society in the UK to offer training courses in New Zealand. These courses aim to educate recreational divers about underwater archaeology, site recording, and conservation. Interest in these courses has grown annually and continues to see the public learning about maritime archaeology.

Recent academic studies have also brought awareness to New Zealand's maritime heritage. Research theses have included some of New Zealand's most famous shipwrecks. One such study investigated two iconic maritime heritage resources: New Zealand's earliest known European shipwreck, *Endeavour*, wrecked 1795, and *Edwin Fox*, preserved in a drydock in Picton (Bennett, 2021). The study compared ship components and technologies to understand the construction and repurposing of British East Indiamen during a period of globalised interactions.

Another notable project that has brought awareness to New Zealand's UCH is the HMS *Buffalo* Re-examination Project. Initiated in late 2020 by Kurt Bennett, Matthew Gainsford, and Rebecca Cox, the project aimed to promote maritime archaeology in New Zealand through the detailed study of the HMS *Buffalo* shipwreck. The project involved conducting an underwater archaeological survey to create a comprehensive site plan and 3D digital model, alongside material sampling to understand the ship's construction. Public outreach was a significant component, with educational programs designed for local schools to engage students in maritime archaeology. This initiative successfully increased community interest and awareness of New Zealand's maritime heritage (Bennett et al., 2021).

While academic and public archaeology projects have significantly contributed to exploring New Zealand's maritime heritage, other sites have been exposed through anthropogenic and natural site formation processes, underscoring the country's legislative framework for UCH protection. For instance, the *Daring* shipwreck, a nineteenth-century schooner uncovered by shifting sands north of Muriwai Beach in 2018, exemplifies this phenomenon. *Daring*, remarkably well-preserved after 153 years, highlighted the application of the *Heritage New Zealand Pouhere Taonga Act 2014*, which serves automatic protection to shipwrecks having wrecked pre-1900 (Bennett et al., 2018). This automatic protection ensures that any modifications or disturbances to an archaeological site are regulated and authorised, thereby ensuring the recovery of information through investigative processes. This, however, did not deter memento seekers and fossickers to the site which contributed to the vessel's rapid degradation.

New Zealand has important UCH in its waters; however, the discipline of maritime archaeology is historically undeveloped. Whilst public groups were established in the 1990s to bring awareness to sites, it has only been in the last decade that professional research has been conducted. As offshore development threatens UCH in New Zealand waters, there will be an increasing need for the nation to invest in establishing important roles at regulatory and educational levels and engage with maritime heritage specialists to manage threats and impacts to sites.

7.4 Legislative Framework

New Zealand has an integrated approach to coastal and maritime heritage management, with an assortment of laws to govern activities, or projects, that occur within this environment and the effects that arise from those activities. Consideration of both activity and effects are tasked to several regulatory bodies, depending on the location of the activity and the type of effects that are generated.

In the context of seabed mining, there are three key pieces of legislation that apply in New Zealand:

- *Crown Minerals Act 1991* (CMA 1991) governs minerals permits for prospecting, exploring and mining Crown-owned minerals.

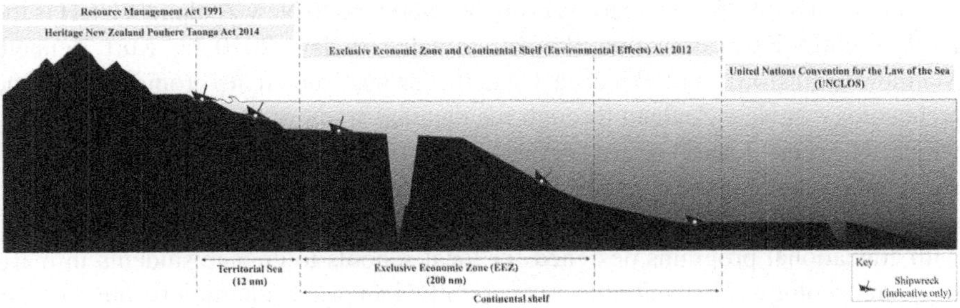

Fig. 7.2 Schematic showing heritage legislation applicable to the marine area. (Source: Authors' own)

- *Resource Management Act 1991* (RMA 1991) regulates the environmental effects arising from seabed mining activities, land-based and occurring in New Zealand's territorial waters (12 nm) through requirement of resource consents to carry out the mining and associated activity.
- *Exclusive Economic Zone and Continental Shelf (Environmental Effects) Act 2012* (EEZ Act 2012) regulates the environmental effects, through marine consents to undertake seabed mining activities in the Exclusive Economic Zone (from 12 nm to 200 nm) (EEZ) and continental shelf.

Any seabed mining activity outside this national jurisdiction is governed by the United Nations Convention on the Law of the Sea (LOSC), and subject to the rules, regulations and procedures of the International Seabed Authority (Fig. 7.2) (see Chaps. 2 and 3, this vol.).

7.4.1 Legislation Managing Historic Heritage

There are two key pieces of legislation operating in tandem to manage and protect historic heritage in New Zealand. The RMA 1991 seeks to avoid, remedy and mitigate any adverse effects on, *inter alia*, historic heritage; and the *Heritage New Zealand Pouhere Taonga Act 2014* (HNZPTA 2014) provides a process to regulate the modification or destruction of an archaeological site.

Historic heritage is a broad term defined in the RMA 1991 (s2) as 'those natural and physical resources that contribute to an understanding and appreciation of New Zealand's history and cultures' including historic buildings, structures, areas, sites of significance to Māori, and of most relevance to this paper, archaeological sites. The RMA 1991 (s5) provides an effects-based regime to regulate any activity that may generate adverse effects on the environment, with the purpose of promoting sustainable management of natural and physical resources. This includes managing the effects on historic heritage, which by its definition, includes archaeological sites. The RMA 1991 (s6(f)) also lists the protection of historic heritage as a matter of national importance and as such any person exercising a function or power under the RMA must 'recognise and provide for' the protection, including when exercising decision-making powers in respect of a resource consent in the coastal marine area.

Archaeological sites are defined in the HNZPTA 2014 (s6) as any place in New Zealand associated with human activity before 1900 or the site of a shipwreck that occurred before 1900 that provides evidence to the history of the country. The HNZPTA 2014 (s43) also provides that a post 1900 site may be declared an archaeological site where its significance allows. Section 87 (HNZPTA 2014) makes it unlawful for anyone to modify or destroy an archaeological site without the prior consent of Heritage New Zealand Pouhere Taonga. This consent comes in the form of an authority which requires reporting in order to collect information and expand the knowledge of New Zealand's past. In effect, preserving in record what cannot be preserved *in situ*.

Outside the territorial waters, the EEZ Act 2012 applies to all activities within the EEZ and continental shelf and provides a regime to manage the effects on the environment of any activity that happens within its jurisdiction. At first blush, the EEZ Act 2012 is similar to the RMA 1991 in its purpose and effects-based approach, however differences in its definitions mean it is applied quite differently.

The EEZ Act 2012 promotes sustainable management of natural resources, which is a narrower term than that of the RMA 1991, where the latter extends to include physical resources—including man-made structures—the EEZ Act does not. The definition for sustainable management also has subtle differences in both acts, however, they are united in the aim of 'avoiding, remedying, or mitigating any adverse effects of activities on the environment' (RMA 1991, s5(2)(c); EEZ Act 2012, s10(2)(c)). The difference is in the definition of environment—where the RMA again includes physical resources, the EEZ Act is limited to only natural resources, under which UCH would not fall. Internationally, there is recognition and protection afforded to UCH through the UNESCO 2001 Convention for the Protection of Underwater Cultural Heritage, however, this has not been ratified by New Zealand.

7.4.2 *Protecting Underwater Cultural Heritage Under the New Zealand Framework*

Whilst not specific to the protection of UCH, the current legal framework in New Zealand does provide some level of recognition and consideration as to adverse effects on heritage that lies in the seabed, within its Territorial Sea. The RMA 1991 and the HNZPTA 2014 both require assessment of any effects generated by seabed mining, which would form part of the overall decision of whether to grant the resource consent (RMA 1991) or archaeological authority (HNZPTA 2014). Notwithstanding these acts providing for UCH, there is not the same consideration in the EEZ Act 2012, and as such there is a level of risk that applies when seabed mining is proposed outside of 12 nm.

7.5 Current State of Deep Sea Underwater Cultural Heritage

New Zealand is home to a variety of historic deep-water shipwreck sites, each with its own unique story. Notable among these are the SS *Ventnor*, RMS *Niagara*, and SS *Port Kembla*, which lie up to and greater than 100 m deep, preserving fragments of maritime history and which have been relocated. These wrecks offer invaluable insights into the past yet remain largely unexplored by professional maritime archaeologists. What all these shipwrecks have in common, however, is that they have been visited by technical divers. The potential for discovery and preservation of these UCH sites is immense, highlighting the need for stakeholder collaboration, dedicated research and protection efforts.

7.5.1 SS Ventnor

In 2014, a team of technical divers explored the wreck of the SS *Ventnor*, which lies at a depth of 147 metres (482 feet) off the Northland coast—making it the deepest dived wreck in New Zealand to date, a remarkable achievement in the field of underwater heritage. The ship, constructed in Scotland in 1901, sank in 1902 while en route from Wellington to Hong Kong, carrying the remains of 499 Chinese miners intended for repatriation to their homeland.

Over a series of dives, artefacts were recovered from the site, including plates, a porthole, and a small bell, to aid in identifying the wreck and later with the intention that they be presented to the Chinese government (Edwards and Jamieson, 2014; Gordon 2022:114 and 173). At the time, the wreck was not legally classified as an archaeological site, and the removal of artefacts was therefore not prohibited. However, the situation sparked public concern—particularly from heritage advocates and Chinese New Zealand communities—due to the cultural and spiritual significance of the site.

In response, Heritage New Zealand Pouhere Taonga designated the wreck as an archaeological site under section 6(b) of the Heritage New Zealand Pouhere Taonga Act, extending legal protection to the site. The artefacts recovered were also classified as protected objects under the Protected Objects Act 1975, meaning that any export would require a permit from the Ministry of Culture and Heritage (Ingram et al., 2007).

The Ventnor case brought renewed attention to the unique challenges of managing deepwater heritage sites, particularly where culturally sensitive material is involved. While the project did not involve formal collaboration between technical divers, archaeologists or heritage agencies, it nonetheless highlights the growing role technical divers can play in accessing and documenting deep, otherwise unreachable sites. Looking ahead, future efforts to manage such locations may benefit from structured partnerships that draw on the complementary expertise of divers, archaeologists, and heritage professionals.

The story of the *Ventnor* offers important insight into the fragility and significance of New Zealand's underwater cultural heritage. Currently, the shipwreck remains at the bottom of the ocean and continues to serve as a reminder of historical events as well as a habitat for marine life. However, it faces threats from deep-sea trawling which can cause significant damage to the wreck and its surrounding ecosystem (see Jarvis, 2024 for further discussion of trawl damage to UCH). Additionally, the potential for deep sea mining, if conducted in the area, poses a risk, as this activity can lead to habitat destruction, sediment plumes, release of toxins and modify archaeological contexts. These activities may threaten the preservation of the shipwreck and the marine biodiversity it supports. Without coordinated management and protective frameworks, historically significant sites, like SS *Ventnor*, risk being disturbed or lost before their stories can be fully understood or shared.

7.5.2 *RMS* Niagara

On 19 June 1940, the RMS *Niagara* encountered a mine laid by the German auxiliary cruiser *Orion* near Whangārei while on its voyage from Auckland to Vancouver. The explosion caused the vessel to sink, and it now rests at a depth of 121 m (396 ft) in the shipping channel between the Hen and Chicken Islands and the Mokohīnau Islands. During World War II, the wreck was salvaged to recover a secret cargo of gold it was carrying. The site was first explored by technical divers in 1999 and most recently in 2024 (Gordon, 2005).

The RMS *Niagara*, like the SS *Ventnor*, does not automatically receive legal protection as an archaeological site because it sank after 1900.[1] The depth of the wreck has limited the number of divers visiting the site, but advancements in technical diving equipment, personal ROVs, and submarines have made deeper sites like the RMS *Niagara* more accessible. Without legal protection and with increasing visitation, the risk of the site being altered or destroyed is growing. Additionally, the RMS *Niagara* has been leaking oil into the surrounding marine environment, complicating its status further (Gordon, 2005). Various groups claim that significant amounts of oil remain in the ship's bunkers, though officials dispute this. Oil leaks are widely reported and criticised by recreational fishers and conservationists. Consequently, the RMS *Niagara* shipwreck poses a challenge for both heritage and environmental management that remains unresolved (see Chap. 6, this vol.).

[1] The RMS *Niagara* wreck site, however, is located in a cable protection area (Submarine Cables and Pipelines Protection Act 1996), so while the site is not afforded heritage-related legal protection, the cable protection area does provide some indirect protection through regulating access to the site.

7.5.3 SS Port Kembla

The SS *Port Kembla*, a steamer operated by the Commonwealth and Dominion Line, was named after the Australian port of the same name. On 18 September 1917, while transporting a cargo of frozen produce, wool, skins, tallow, jams, and lead from Australia to the United Kingdom, the ship struck a mine laid by the German auxiliary cruiser SMS *Wolf* near Farewell Spit, New Zealand. The explosion caused the vessel to sink quickly, but all 60 crew members were rescued by the steamer *Regulus* and safely brought to Nelson (Hindmarsh, 2021).

Today, the wreck of the SS *Port Kembla* rests at a depth of 96 m (315 ft) and has become a popular site for wreck diving. Divers are attracted to the site for its historical significance and the chance to explore a well-preserved World War I-era shipwreck. The wreck offers a unique window into early twentieth-century maritime history and the dangers faced by ships during wartime.

7.5.4 UCH Summary

In summary, New Zealand's UCH is rich and diverse, encompassing numerous shipwrecks from both World War I and World War II. These military and naval vessels, now resting on the ocean floor, offer invaluable insights into historical naval operations and wartime activities. The cold, deep waters around New Zealand have also contributed to the good state of preservation of these wrecks, providing a unique window into the past.

In addition to military vessels, New Zealand's waters also hold the remains of other nation's vessels. These wrecks reflect a shared maritime heritage, representing the era of exploration, trade, and colonial expansion. The preservation of these sites is crucial for understanding and appreciating the complex history they represent. Together, these military and foreign shipwrecks form an important part of New Zealand's UCH, offering significant potential for both historical research and heritage tourism.

The current situation of deep-sea wrecks in New Zealand waters remains largely unexplored by professional maritime archaeologists. Despite the presence of significant wrecks such as the SS *Ventnor*, RMS *Niagara*, and SS *Port Kembla*, and having been visited by technical divers, no comprehensive professional maritime archaeological surveys or investigations have been conducted. This gap in professional exploration leaves many questions about the historical and environmental impacts of these wrecks unanswered. The absence of detailed surveys underscores the need for dedicated maritime archaeological efforts to better understand and preserve these underwater cultural heritage sites. Additionally, the increasing interest in deep sea mining activities further highlights the urgency for thorough investigations to ensure that these wrecks are not inadvertently damaged or destroyed.

7.6 Potential Threats

Deep sea mining poses a significant potential threat to UCH, particularly in New Zealand where many shipwreck sites remain largely unexplored or are yet to be relocated. While technical divers have accessed sites between 100 and 200 m, no professional maritime archaeological surveys have been completed, leaving a gap in our understanding of these underwater resources. The potentially unknown UCH resources at depths greater than 200 m could hold invaluable historical and cultural information. The scarcity of these sites potentially increases their significance, as each site could provide unique insights into New Zealand's maritime history and heritage. This is especially important considering that the deeper sites, though fewer in number, may contain well-preserved artefacts due to the reduced human activity and slower degradation processes at such depths (UNESCO, 2023).

Existing nearshore sand mining activities have already demonstrated the potential for significant environmental and cultural damage. For instance, the Taranaki coast, known for its rich maritime history, faces threats from proposed sand mining operations. The PS *Tasmanian Maid* wreck site, a well-known historical site along this coast, exemplifies the potential risks. *Tasmanian Maid*, a paddle steamer that sank in 1868, is a significant part of New Zealand's maritime heritage (Heritage New Zealand, 2024). Coastal sand mining could exacerbate erosion and disturb these underwater sites, leading to the loss of invaluable historical information. Additionally, other historically significant ships, such as the HMS *Orpheus*, could be at risk from coastal sand mining or dredging activities if conducted in the same area. *Orpheus*, which sank in 1863, remains New Zealand's worst maritime disaster with 189 lives lost (NZHistory, 2020). These nearshore examples highlight how human activities have the potential to impact UCH, underscoring the possible threats that deep sea mining could pose to deeper, less explored sites. In the New Zealand context, the preservation of these sites is crucial not only for their historical value but also for their role in understanding the broader maritime history of the region. The loss of these sites would mean a significant gap in the historical narrative of New Zealand's maritime past.

To mitigate these threats, it is essential to invest in maritime archaeology programs and employ maritime archaeologists at regulatory and policy levels. Whilst, educating the public and stakeholders about the importance of UCH can foster greater appreciation and support for preservation efforts. Programs like the introductory courses offered by the Australasian Institute for Maritime Archaeology are steps in the right direction (Australasian Institute for Maritime Archaeology, 2024). Additionally, cross-collaboration between groups such as the National Institute of Water and Atmospheric Research (NIWA) and the Royal New Zealand Navy can be instrumental. Utilising their hydrographic expertise and resources to conduct in-water surveys in collaboration with maritime archaeologists can lead to comprehensive documentation and preservation efforts. For example, the collaboration on the *Buffalo* wreck site with the Navy demonstrated the effectiveness of such interdisciplinary projects (NZ Defence Force, 2024). By conducting professional archaeological surveys and research, we can better document and protect these sites, ensuring that New Zealand's UCH is preserved for future generations. This proactive approach will help safeguard the few but highly significant UCH sites that exist, highlighting their importance in the national and global context.

The significance of having a potential shipwreck in New Zealand waters that shares heritage with other nations is considerable. Such a vessel may be a sovereign vessel, meaning New Zealand, under best practice, should care for another nation's UCH. This approach is underscored by the UNESCO 2001 Convention on the Protection of the Underwater Cultural Heritage, which emphasises the importance of preserving UCH for humanity and encourages international cooperation. Shared heritage sites can foster cross-collaboration between nations, leading to joint research projects, shared resources, and a deeper mutual understanding of maritime history. This collaboration can also bring economic benefits through tourism and educational programs, further highlighting the importance of preserving these underwater cultural treasures.

7.7 Conclusion

New Zealand's deep sea is rich in biodiversity and holds significant UCH. The legislative framework protecting these sites, however, is limited in its coverage and application. While the threat of deep sea mining to New Zealand's UCH is significant, it also presents an opportunity to highlight the importance of such UCH sites. The limited number of known sites at greater depths makes each one even more valuable, necessitating urgent and comprehensive efforts to explore, document, and protect them along with others. The potential erosion of nearshore sites like those along the Taranaki coast and the risks to other wrecks such as the HMS *Orpheus* underscore the need for immediate action and to give serious consideration to other sites in deeper waters. By investing in maritime archaeology and public education, New Zealand can ensure that its UCH is not only preserved but also celebrated for its historical and cultural significance.

References

Australasian Institute for Maritime Archaeology. (2024). *Training courses*. [online] Accessed October 10, 2024, from https://aima.org.au/training-courses.

Bennett, K., Gainsford, M., & Cox, R. (2021). For community, by community—Promoting maritime archaeology in Aotearoa/New Zealand through the re-examination of the HMS Buffalo Shipwreck (1840). *Australasian Journal of Maritime Archaeology, 45*(1), 25–48.

Bennett, K. (2021). *Shipwright artistry: Cultural transmission of British colonial ship design and construction during the eighteenth and nineteenth centuries*. PhD dissertation, College of Humanities, Arts and Social Sciences, Flinders University, Adelaide.

Bennett, K., McIvor, I. H., & Paul, L. (2018). Report on the rescue and preservation of daring, a mid-19th-century schooner built in Aotearoa/New Zealand. *Australasian Journal of Maritime Archaeology, 42*(1), 75–80.

Bennett, K., & Parslow, B. (2025). Between two tides: Managing intertidal shipwrecks as an archaeological resource in Aotearoa New Zealand. In J. Jones, C. Mires, & D. Zwick (Eds.), *The intertidal shipwrecks: Management of a historic resource in an unmanageable environment*. University of Florida Press.

Carter, M. (2019). *Manufacturing identity: Pākehā shipbuilding and the 'middle ground' in Aotearoa New Zealand*. PhD Thesis, La Trobe University, Melbourne.

Carter, M., & Bennett, K. (2022). Underwater cultural heritage in Aotearoa New Zealand: Challenges and opportunities. In A. Hafner, H. Öniz, L. Semaan, & C. J. Underwood (Eds.), *Heritage under water at risk: Treats – challenges – solutions* (pp. 94–98). ICOMOS.

Carter, M., & Dodd, A. (2015). Maritime archaeology in New Zealand: Trials, tribulations and opportunities. In S. Tripati (Ed.), *Shipwrecks around the world; Revelations of the past*. Delta Book World.

Churchill, D. (1991). The Maritime Archaeological Association of New Zealand MAANZ (Inc). *Bulletin of the Australasian Institute for Maritime Archaeology 15*(1), 7–10.

Davies, P., & Myburgh, P. (2006). New Zealand. In S. Dromgoole (Ed.), *The protection of the underwater cultural heritage: National perspectives in light of the UNESCO Convention 2001* (pp. 189–215). Martinus Nijhoff Publishers.

Department of Conservation. (2023). *New Zealand's marine biodiversity*. Accessed October 2, 2024, from https://www.doc.govt.nz/nature/habitats/marine/new-zealands-marine-biodiversity/

Dodd, A. (2003). Opportunities for underwater archaeology in New Zealand. *Archaeology in New Zealand, 46*(3), 151–160.

Edwards, B., & Jamieson, B. (2014). *Archaeological assessment of the SS Ventnor ship wreck Hokianga Harbour mouth*. NZAA Site No: O05/350. Report prepared for the Northland Regional Council.

GNS Science. (2023). *Geology of New Zealand*. Accessed October 2, 2024, from https://www.gns.cri.nz/our-science/land-and-marine-geoscience/te-riu-a-maui-our-continent/geology-of-new-zealand/.

Gordon, K. (2005). *Deep water gold: The story of RMS Niagara, the quest for New Zealand's greatest shipwreck treasure*. Searov Technologies.

Gordon, K. (2022) *SS Ventnor: Ghost ship of the Hokianga*. Searov Publications: Whangarei.

Heritage New Zealand. (2024). *PS Tasmanian maid wreck site*. [online] Accessed October 10, 2024, from https://www.heritage.org.nz/list-details/9521/PS Tasmanian Maid Wreck Site.

Hindmarsh, G. (2021). Ship went down in 20 minutes as German raider's mine struck off Farewell Spit. Nelson Mail. Accessed November 9, 2024, from https://www.stuff.co.nz/nelson-mail/news/125791771/ship-went-down-in-20-minutes-as-german-raiders-mine-struck-off-farewell-spit.

Ingram, C. W., Wheatley, P. O., Diggle, L., Diggle, E., & Gordon, K. (2007). *New Zealand shipwrecks: Over 200 years of disasters at sea*. Hodder Moa.

International Union for Conservation of Nature (IUCN). (2022). *Deep-sea mining: An update*. [PDF] Accessed October 5, 2024, from https://iucn.org/sites/default/files/2022-07/iucn-issues-brief_dsm_update_final.pdf.

Jarvis, C. (2024). *Threats to our ocean heritage: Bottom trawling*. Springer.

Johns, D. A., Irwin, G. J., & Sung, Y. K. (2014). An early sophisticated East Polynesian voyaging canoe discovered on New Zealand's coast. *PNAS, 111*(41), 14728–14733.

Linley, T. D., Stewart, A. L., McMillan, P. J., Clark, M. R., Gerringer, M. E., Drazen, J. C., Fujii, T., & Jamieson, A. J. (2017). Bait attending fishes of the Abyssal zone and Hadal boundary: Community structure, functional groups and species distribution in the Kermadec, New Hebrides and Mariana trenches. *Deep Sea Research Part I: Oceanographic Research Papers, 121*, 38–53. Accessed November 9, 2024, from https://doi.org/10.1016/j.dsr.2016.12.009

Mining Digital. (2023). *New Zealand seabed mining moves closer with EPA approval*. Accessed October 2, 2024, from https://miningdigital.com/supply-chain-and-operations/new-zealand-seabed-mining-moves-closer-epa-approval.
NIWA. (2023). *New map reveals New Zealand's seafloor in stunning detail*. Accessed October 2, 2024, from https://niwa.co.nz/news/new-map-reveals-new-zealands-seafloor-stunning-detail.
NZ Defence Force. (2024). Things that go beep in the estuary. *Navy Today Issue 265*. [online] Accessed October 10, 2024, from https://issuu.com/nzdefenceforce/docs/navytoday_issue265.
NZHistory. (2020). *New Zealand's worst shipwreck*. [online] Accessed October 10, 2024, from https://nzhistory.govt.nz/sinking-of-hms-em-orpheus-em-nzs-worst-shipwreck.
Smith, I. (2019). Pākehā Settlements in a Māori World. Bridget Williams Books, Wellington.
Te Ara. (2023a). *Sea floor geology*. Accessed October 2, 2024, from https://teara.govt.nz/en/sea-floor-geology.
Te Ara. (2023b). *Marine minerals*. Accessed October 10, 2024, from https://teara.govt.nz/en/marine-minerals.
UNESCO. (2023). *Threats to the underwater cultural heritage*. [online] Accessed October 10, 2024, from https://www.unesco.org/en/underwater-heritage/threats.

Chapter 8
Moana Matters: Culture and Deep Sea Mining in Oceania

Elise Huffer and Renee Hill-Lewenilovo

Abstract Cultural connectivity has been a marker of Oceania since time immemorial. Islands were fished out of the water by ancestral demi-gods; sea routes were mapped and shared with kindred animals; currents, winds, and stars were felt, breathed in, and drawn in sand and in minds. Sharks, dolphins, and turtles are still called today across different islands Place names are shared across vast expanses and ancient sacred complexes unite communities from different corners of Oceania. Deep sea mining (DSM) threatens this long standing connectivity, compounding the already serious impacts of climate change which are damaging the ocean and islands.

Keywords Deep Sea Mining (DSM) · Oceania · Cultural connectivity · Traditional ecological knowledge · Spiritual turmoil

> Oceania is vast, Oceania is expanding, Oceania is hospitable and generous, Oceania is humanity rising from the depths of brine and regions of fire deeper still, Oceania is us. We are the sea, we are the ocean…
>
> —Epeli Hau'ofa, We Are the Ocean: Selected Works

E. Huffer (✉)
University of the South Pacific, Suva, Fiji

R. Hill-Lewenilovo
Ocean Voices Fellow (2024–2025), Suva, Fiji

C. Jarvis (ed.), *Threats to Our Ocean Heritage: Deep Sea Mining*,
SpringerBriefs in Archaeology, https://doi.org/10.1007/978-3-031-98238-5_8

8.1 Introduction

Epeli Hau'ofa's often quoted 'we are the sea, we are the ocean', affirms the depth of connection of the people of Oceania[1] with the ocean. Cultural connectivity has been a marker of Oceania since time immemorial. Islands were fished out of the water by ancestral demi-gods; sea routes were mapped and shared with kindred animals; currents, winds, and stars were felt, breathed in, and drawn in sand and in minds. Sharks, dolphins, and turtles are still called[2] today across different islands (Torrente et al., 2018). Place names are shared across vast expanses and ancient sacred complexes unite communities from different corners of Oceania. Deep sea mining (DSM) threatens this long-standing connectivity, compounding the already serious impacts of climate change which are damaging the ocean and islands. As the region and global community collectively confront DSM, to understand the cultural perspective we turn first to Oceanian scholars, activists, artists, navigators, cultural experts, theologians, and practitioners for guidance. They are telling stories that explain and reinforce traditional bonds to the moana. They provide accounts of kinship and retell ocean histories, genealogies, and contemporary navigational feats. There are many to name and to listen to.

A few examples include Tongan academic Tēvita 'O Ka'ili, whose work on Tongan genealogy illustrates the intimate and enduring relationships with ocean ancestors and the relevance of these to ecological stewardship (Ka'ili, 2024; *Lala's LALINOA*, 2024); Māori scholar Dan Hikuroa, whose writing on mātauranga Māori and kaitiatikanga demonstrates how the principles and values that underpin them, are key to understanding the depth of caring for land and ocean, which go beyond but are compatible with Western scientific approaches and can help us better manage the ocean (and land) (McAllister et al., 2023; Hikura, 2020); Uncle Sol (Solomon Pili Kaho'ohalahala) whose powerful voice at the International Seabed Authority (ISA) and in other fora urges leaders and participants to focus on culture and to consider the potential impacts of seabed mining on cultural connections and heritage (see Chap. 4, this vol.; Segalov, 2023; Pili Kaho'ohalahala, 2024); and scholar 'Atu Emberson-Bain with her sons (Siale and Menino Bain-Vete), daughter (Anga'aefonu (Fonu) Bain-Vete) and grandchildren (Tahi and Mahina Bain-Vete) who received the 2024 UK Rubery Book of the Year Award for their children's and

[1]The term Oceania refers to all the Pacific Islands and includes Aotearoa/New Zealand and Australia. Although the term was initially coined during colonial times it has been reclaimed (and popularised through the work of Epeli Hau'ofa) to emphasise the connections between Pacific peoples across the vast ocean, past and present, and to break away from narratives of smallness, isolation and vulnerability imposed by Western perspectives of the Pacific Islands.

[2]Different communities across Oceania have traditions of calling sea creatures using a range of techniques and rituals. The calling can be done for fishing or other purposes.

young adults' book *Cries from the Moana* (2023), about the dangers of deep sea mining, and provide a testament to the tradition and potency of intergenerational storytelling including in print (Fiji One News Team, 2024). These voices echo those of coastal and ocean community members whose lives are centred around the sea, some of whom could experience the direct threats of DSM.

As noted by Julian Arato et al. (2024) and Uncle Sol,[3] culture has not featured in the global discussions about deep sea mining, even though it is a central concern of communities in different parts of the world (DOSI, 2021). Connections to the ocean may be historical, spiritual, ancestral, practical and/or use based, but they are generally a combination of all, especially for communities in Oceania. Arato and colleagues raise the issue of the distinction between the use-value and the identity (sacred, spiritual, ancestral, genealogical) value of the ocean held by communities as posing both challenges and opportunities in the context of deep sea mining and the BBNJ Agreement (Arato et al., 2024). A challenge they suggest is that the use value might take away from the identity value by permitting room for negotiation rather than 'allow[ing] for a wholesale rejection of competing claims… and … prevent[ing] most other types of engagement with the deep seabed' (p.69) at least in the context of international law.

Although, as stated, the authors are referring to this distinction from an international law viewpoint in a discussion on Indigenous approaches in areas beyond national jurisdictions, it may also be a useful prism through which to consider the current opposing views about DSM in the Pacific. The views split between those fighting against it from a cultural values and livelihood perspective (which are intimately linked) and those, particularly the pro-DSM governments of Papua New Guinea, Cook Islands, Nauru, Kiribati, and Tonga (along with their associated mining company), who are convinced of the economic benefits of DSM. Their focus on use-values of the ocean tend to negate or, at best, ignore the identity values put forward by communities. This is not altogether surprising as use-values generally permeate government ocean policy in the Pacific in spite of statements about culture.[4] In the case of DSM, this is amplified by Pacific states' legal ownership of seabed mineral resources within Economic Exclusive Zones (EEZ)—a form of ownership which effectively denies communities from any benefit sharing, while leaving them with the burden of potential impact.

[3] Uncle Sol states, 'When I challenged The Metals Company on the absence of cultural considerations in their deep sea mining proposal presentation, they answered that they hadn't considered it prior to me coming to the ISA. They say they now plan to address this issue, but still do not know how to do so'. See: Pili Kaho'ohalahala (2024).

[4] Vanuatu is the only Pacific country which has placed culture at the foundation of its ocean policy. It is also strongly opposed to DSM.

8.2 EEZs of the Pacific

8.2.1 Papua New Guinea

Conflict between pro-DSM governments and communities, particularly in the case of prospecting[5] within national EEZs, has been an ongoing feature of DSM in Oceania, where communities view culture and environment as one and central to their lives for both identity and (sustainable) livelihood.[6] This is exemplified by the case of Papua New Guinea with the Solwara 1 prospecting's first phase in 2011. Solwara 1 was Oceania's first attempted foray into DSM, with prospecting taking place in the Bismarck Sea between the islands of New Ireland and New Britain. This first attempt failed, as Nautilus Minerals, the Canadian mining company which had partnered with the PNG government, declared bankruptcy in 2019. Opposing communities and civil society established the Alliance of Solwara Warriors in 2016, a group which has aimed to put a complete stop to DSM in PNG and the Pacific citing cultural and livelihood values and practices. Prospecting is now on the rebound in the Bismarck Sea with new activity underway via Deep Sea Mining Finance (DSMF), the company which took over Nautilus Minerals in 2019. The PNG government has seemingly approved this new prospecting effort in a bid to reverse the financial losses (around USD125 million) incurred through its partnership with Nautilus Minerals.[7] The venture is once again threatening the cultural wellbeing of neighbouring communities as illustrated by a PBS NewsHour broadcast series of interviews. In one of these, Chris Malgan, a traditional shark caller and chief of Kono village in New Ireland (located near where the prospecting is taking place), states, 'The ocean is our life. For us, the coastal people, it is our garden. We catch fish and share with others in the village without being paid. When we catch fish and sharks, we share with other families who need it to spread love and kindness' (Marx & Kiernan, 2024b). What Malgan and other community members fear is the destruction of their entire way of life, which is already being impacted by logging and the coastal impacts of climate change. DSM is the nail in the coffin: 'If this project goes ahead, the coastal communities won't be able to catch fish or shark anymore... and we will lose our culture that has been maintained for centuries' (Kessler, 2024).

'In our culture, a shark is seen as a blessing': the connection with sharks and shark calling discussed above is not unique to New Ireland's coastal communities, as shown through the recently published story of Uein Buranibwe and Temaei

[5] Prospecting in deep sea mining refers to the search for minerals which includes the use of machinery and equipment creating disturbance at sea.

[6] As Uncle Sol states, 'In our culture, there is no separation between us and nature—it is who we are. No matter where we are in the deepest part of the ocean, the surface of the waters, to the land and the mountains and the sky; we are part of all of it. So when industries are plotting to destroy the area of creation and have no regard for that part of our identity, it hurts'. See Pili Kahoʻohalahala (2024).

[7] A PBS NewsHour segment, the third and last in the series, raises doubts about the approval process within the PNG government. See: Marx and Kiernan (2024a).

Tontaake (Boucher et al., 2024). Stranded at sea for 33 days, the fishermen from Kiribati lassoed a shark to guide their small boat to Namdrik Atoll in the Marshall Islands. It is through the memory of his father's stories about using sharks to 'pull his canoe back to shore' when returning home from fishing, that Uein recalled that 'sharks like to swim towards land'. Upon reaching Namdrik, the fishermen discovered a genealogical link—Temaei's uncle, lost at sea decades earlier, had settled there, forming a new family branch. These connections, both physical and spiritual, reveal the unifying role of sharks and other marine species/ancestors as they connect Oceania's people, cultures, narratives and histories, tying together navigation, survival, and kinship across the Pacific.

8.2.2 *Ecology and Culture*

The cultural connection with sharks is intertwined with their ecological and geological significance. Shark teeth, often found on seamounts, can serve as nuclei for polymetallic nodules and cobalt-rich crusts, linking these iconic creatures to the frontier of deep sea mining. For Māori, the *pounamu* or greenstone of Aotearoa/New Zealand originate as soft matter inside a shark, hardening upon exposure to air (Nichols, 1992).[8] The geological connection between shark teeth, whale bones, crustacean shells, and ore formation reveals an intricate interplay of marine biodiversity, oceanic processes, and cultural traditions—an interplay now threatened by DSM. In Oceania, sharks are not just keystone species but cultural icons, embodying both heritage and ecological vitality. Acting as navigators, ecosystem indicators, and geological connectors, their presence on seamounts signifies the health of these marine habitats. This triple role of Ocean Heritage—ecological, geological and cultural—amplifies the urgency of protecting seamounts and the species they sustain.

In her rich discussion of the 'sea of voices' of stakeholders in PNG during the initial Solwara 1 experiment with Nautilus, Haley Phillips (2019) highlights the discordance between the economic use perspective put forward by the government and Nautilus, and the community position which draws on an 'ocean-based lifeworld'.[9] She demonstrates how the cultural relationship of communities with the ocean is negated by both Nautilus and the government through a range of

[8] This report chapter on sharks examines the cultural and ecological significance of sharks across the South Pacific, highlighting their roles in indigenous folklore and as keystone species in marine ecosystems. Nichols discusses traditional beliefs, such as the worship of sharks as ancestral spirits in Melanesian islands like the Solomon Islands and Papua New Guinea, and the Fijian legend of Dakuwaqa, a mythical shark deity. The report underscores the importance of integrating cultural perspectives into conservation efforts to ensure the sustainable management of shark populations in the region.

[9] Phillips draws on Ingersoll's (2016) notion of 'seascape epistemology', which is 'an approach to knowing presumed on a knowledge of the sea, which tells one how to move through it, how to approach life and knowing through the movements of the world. It is an approach to knowing through a visual, spiritual, intellectual, and embodied literacy of land and sea' (Ingersoll, 2016, pp. 5–6).

tactics, including a dismissive Environmental Impact Statement and shoddy consultation process, statements to the effect that local people are not connected to the sea beyond the shore, and reaffirmations of state power over the ocean seabed. She posits that 'like Nautilus, the national government has much to gain by refusing to acknowledge any cultural connection its citizens have to the Solwara1 site' (Phillips, 2019, p. 13) as it seeks to legitimise its own economic interests and control of seabed resources. This phenomenon is also described by John Childs in his assessment of Solwara 1 (Childs, 2019). He notably illustrates how the 'distancing of the deep ocean' from people and culture by the proponents of DSM, clashes with the cosmological world of coastal communities. Beyond the environmental impacts, Childs describes how DSM risks disturbing the spirits, the 'masalai—some of which are understood as guardians of the seabed and its resources' (Childs, 2019). Ignoring this factor, he argues, may not only marginalise the communities but also lead to 'spiritual turmoil'. He further notes how the removal of minerals, which, from the perspective of local communities, have been created by spiritual beings, is problematic: 'Thus, as copper "resurfaces" in the objects and technologies of the future—in batteries and wiring—it also carries a spirituality from the region where it originated'. The arguments put forward by governments and mining companies that DSM will have fewer impacts on communities than land-based mining, seem, in light of the world views of these communities, misguided and disingenuous. Acknowledging these cultural relationships would lead to a more balanced approach to DSM but the willingness to do this is currently absent.[10]

The profound spiritual turmoil caused by resource extraction is powerfully articulated by Katerina Martina Teaiwa in *Consuming Ocean Island* (2014). Teaiwa illustrates how the spirit of Banaba was extracted alongside its phosphate, fuelling global agricultural production while leaving the Banabans displaced and spiritually unmoored.[11] This echoes the experiences of coastal communities affected by DSM, where the disruption of *masalai*—spiritual guardians of the seabed—creates a similar cosmological imbalance. In both cases, the commodification of sacred elements for global gain undermines the intricate relationships between people and their environments. For the Banabans, their land (*te aba*) was not merely a resource but a spiritual and cultural cornerstone; its loss resulted in profound grief and cultural dispossession. Similarly, in Papua New Guinea, the removal of seabed minerals is seen as extracting not just physical wealth but spiritual essence, alienating communities from their cosmological worldviews. Ignoring these spiritual dimensions

[10] In the case of Papua New Guinea, the lack of transparency around the current Solwara 1 phase, as documented by PBS NewsHour does not augur well for communities. See: Marx and Kiernan (2024a).

[11] This work explores the cultural, environmental, and spiritual impacts of phosphate mining on Banaba, highlighting how the extraction of phosphate not only fuelled global agricultural production but also displaced the Banaban people and severed their connection to their homeland. Teaiwa critiques the commodification of sacred land and emphasizes the ongoing consequences of colonial resource extraction for Indigenous communities.

perpetuates marginalisation and cultural erasure, highlighting the urgent need for governance models that honour and incorporate Indigenous cosmologies.

8.2.3 Food, Culture, and the Environment

The intimate relationship between people, food, and the environment further illustrates the harm of extraction. On Banaba and Nauru, phosphate mining fuelled agricultural growth in nations like Australia, New Zealand, and Japan, but at a devastating cost. While the fertile soils of distant lands prospered, Banabans were left with barren landscapes and displaced communities. The voices of Banabans were silenced as British, Australian, and New Zealand investors prioritised profits, embodying the deep inequities of extractive industries where distant gains are built on local suffering.

In Fiji, the wisdom passed down through generations about fish poisoning (*kana ni ika dauveve*) offers a powerful example of how Pacific communities have long navigated environmental challenges. Fijian Traditional Ecological Knowledge (TEK) embodies a deep connection to the sea and its rhythms, guiding communities in protecting their food systems and avoiding harm. This knowledge includes identifying toxic fish species, understanding seasonal patterns tied to the spawning of *balolo* (edible seaworms), and recognising specific reefs as danger zones. For instance, reefs in Kadavu and Lau, marked by abundant *bulewa* (soft coral), are known as hotspots for toxic fish. Folk tests and traditional remedies, preserved through storytelling and practice, demonstrate the resourcefulness and adaptability of Fijians in responding to these risks. By blending wisdom with resilience, these practices sustain a harmonious balance with the ocean, reflecting a relationship that values respect and coexistence over exploitation.

This profound connection to the marine environment highlights the critical role of cultural knowledge in safeguarding food systems and living in harmony with nature. As the Pacific faces mounting threats like DSM and climate change, the lessons embedded in TEK offer a vital roadmap for resilience. As emphasised in Lako et al., 2023, fish poisoning remains a growing concern in Fiji, exacerbated by rising sea temperatures, cyclone seasons, and environmental disruptions.[12] Heat-resistant toxins such as ciguatoxins, palytoxins, and tetrodotoxins pose serious risks to food safety and public health, challenging conventional mitigation methods. TEK provides essential strategies to navigate these challenges, including identifying toxic hotspots like the Senimuna reef in Kadavu and Kabara reef in Lau, recognising seasonal toxicity patterns during balolo spawning, and employing folk tests to detect danger.

[12] The study underscores how disruptions in marine environments impact food safety and public health. It also emphasises the importance of blending traditional knowledge with scientific approaches to safeguard food systems in the face of environmental challenges.

Interestingly, Lako and colleagues also highlight the role of land-based activities in exacerbating marine toxicity. Copper mining near Udu Point in Vanua Levu and mahogany plantations on Cicia Island in the Lau Group are linked to a higher incidence of fish poisoning in those areas. Interviewees recounted a tragic incident in the 1990s where two individuals died after consuming turtle meat from a village near the mining site, suggesting a potential connection between mining activities and the contamination of marine life. Many Fijians believe that disruptions to the intricate relationships between the environment, land, and people can trigger fish poisoning, reflecting the holistic perspective embedded in TEK.

This worldview underscores the interconnectedness of human actions and natural systems, reminding us of the need to honour and integrate such knowledge to foster ecological balance and community well-being. These adaptive practices illustrate the enduring relationship between Fijians and their marine environment, highlighting the strength of cultural knowledge in addressing modern environmental challenges. By valuing and integrating TEK, Pacific communities can preserve their food systems, maintain ecological balance, and sustain their cultural identity in the face of an uncertain future.

This resilience is not new; historical parallels can be drawn to Polynesian migration, where environmental factors like ciguatera fish poisoning likely compelled communities to make critical survival decisions, as explored in Teina Rongo's research (Rongo et al., 2009).[13] Just as Polynesians relied on navigation and a deep understanding of their environment to escape regions of toxicity and relocate to areas of abundance, Fijians today draw upon TEK to navigate modern challenges. Preventive measures, such as folk tests and the removal of toxic organs, alongside herbal treatments, reflect the same ingenuity and adaptability that have long safeguarded Pacific food systems in the face of ecological disruptions.

Western science highlights the critical risks associated with DSM. Drazen et al. (2020) underscore the potential harm posed by sediment plumes and noise to midwater organisms and essential ecological processes, such as carbon sequestration and nutrient cycling. These processes are vital for maintaining global and regional environmental health, yet their potential disruption by DSM remains underexplored in dialogues with Indigenous Peoples and Local Communities (IPLCs). Bridging this gap is crucial, as DSM directly threatens food systems and ecological stability in the Pacific. Sediment plumes not only disrupt marine habitats and alter food webs but also endanger migratory species that are integral to Pacific Islanders' diets, livelihoods, and cultural identities.

These threats echo the broader vulnerabilities seen in historical cases such as Banaba and Nauru, where resource extraction led to environmental degradation, displacement, and cultural loss. Even Polynesian migration—driven in part by ecological crises—offers lessons about resilience and adaptation in the face of environmental challenges. These historical and ongoing examples highlight the urgency of

[13] See also how the article includes a detailed map of Polynesian migration routes, offering valuable insights into navigation, food security, and the historical connectivity of oceanic communities near regions like the Clarion-Clipperton Zone.

integrating traditional ecological knowledge, respecting Indigenous rights, and adopting precautionary governance frameworks. Such approaches are essential to balancing sustainable development with the preservation of cultural and ecological integrity, ensuring that IPLCs remain active stewards of their oceanic 'sea of islands'.

8.2.4 Cook Islands

DSM in national waters of the Cook Islands is also a profoundly divisive issue. In 2022, the government issued exploration licenses to three mining companies, one of which it co-owns and the Cook Islands Seabed Minerals Authority has established DSM regulations (Cook Islands News, 2023). As the 52nd Underwater Mineral Resources (UMR) Conference, 'the world's largest gathering of ocean mineral stakeholders', was about to take place in Rarotonga in September 2024, a poll conducted by Te Puna Vai Marama (the Cook Islands Centre for Research), found that '85% of respondents agreed that the deep ocean held cultural and spiritual significance for Cook Islanders' and that a majority of respondents were opposed to DSM exploration with 64% opposed to the mining itself. Among all respondents, 64% were against DSM on the basis that the 'deep ocean has cultural and spiritual significance for Cook Islanders' (Tuari'i & Worth, 2024, p. 9).

This poll, conducted online in both Cook Islands Māori and English, marked the first time the public was directly asked their opinion on DSM. Many opposing DSM feel that there has not been a robust consultation about mining in the country and that the mining sector has infiltrated Cook Island's society and economy. An example of this is the funding of many activities including, in what could be seen as a cynical move, the annual Cook Islands cultural festival, the Te Maeva Nui, and other cultural initiatives including an immersion Cook Islands Māori pre-school (Fuatai, 2024). As Louisa Castledine, who hails from Manihiki, which is one of the areas in Cook Islands with the richest mineral deposits, states '… the industry has secured the support of some of our foremost cultural and spiritual leaders by funding their organisations or initiatives' (Fuatai, 2024). In response to this co-opting of culture, a group of individuals, CSOs (including conservation organisations such as Te Ipukarea and Kōrero O Te 'Ōrau, and Te Manava Ora cultural group) and businesses, and with Casteldine and prominent cultural and environmental figures such as Dr Teina Rongo, formed Ocean Ancestors and staged the At Sea, Sail Out Rally (Protect our Ocean: Think Deeply) protest in Rarotonga during the UMR conference. The rally took place on the water with the traditional double hulled canoe, the *Vaka Marumaru Atua*, paddling canoes and other boats as well as on land. Whales, including a mother and her calf, came to the ocean area and right into the harbour including alongside a mining exploration vessel when the Ocean Ancestors were preparing for their rally. The organisers consider this a strong sign of support and mutual recognition: 'that's our ancestors visiting us'.[14]

The rally reflects the deep-rooted connection between the people and their marine environment, a relationship that goes beyond material resources to encompass cultural identity, spiritual values, and traditional practices. The Manihiki Plateau, an oceanic plateau, exemplifies this profound relationship. Formed through volcanic activity more than 120 million years ago, its unique geological features, including seamounts, ridges, and knolls, support fragile ecosystems with sediment-eating organisms, sponges, and other marine species. Its proposed designation as an Ecologically or Biologically Significant Marine Area (EBSA) underscores its importance as a biodiverse yet vulnerable environment.

The plateau's ecological richness stands in stark contrast to the risks posed by deep sea mining, which targets its polymetallic nodules rich in manganese, nickel, cobalt, and other metals. Mining activities threaten to disrupt the seafloor, release harmful pollutants, and destroy habitats that remain poorly understood, with consequences that could reverberate throughout the marine food web. These risks have sparked concern among Cook Islanders, including residents of Manihiki, who argue that mining outside undermines local values. For the Manihiki community, the ocean is a source of spiritual connection, cultural identity, and livelihood, and protecting it is integral to ensuring the well-being of future generations.

Castledine, the spokesperson for Ocean Ancestors, argues that DSM serves external interests and is driven by overconsumption and capitalism. She emphasises that the Cook Islands are already facing the climate crisis, and that mining would further jeopardise their environment and way of life. The Cook Islands have a strong tradition of sustainable resource management, exemplified by practices like *ra'ui* and *tāporoporo*. Castledine believes that the extractive nature of deep sea mining would disrupt these traditional practices and negatively impact the natural resources upon which Cook Islanders depend. Castledine criticises the government's consultation process, arguing that it has been inadequate and lacks clarity regarding the potential impacts of deep sea mining (Fuatai, 2024). She recounts how her own family, despite attending government meetings, still has limited understanding of the environmental and social consequences. This lack of information is particularly concerning given the presence of a significant polymetallic nodule deposit near Manihiki, making it a prime target for mining companies. Castledine points to the tactics employed by the DSM industry, which she describes as a form of neo-colonialism. The strong emphasis on family relationships and collective cultural values, she explains, is being manipulated by companies who are seeking to gain a foothold in the islands. She argues that the mining industry is leveraging the Cook Islands' social structure to their advantage. Castledine highlights the use of euphemisms like 'harvesting minerals' and 'seabed minerals development' by the government and mining advocates to

[14] Pers. comm. June Hosking, Executive President, Te Ipukarea Society (03/12/2024).

downplay the negative connotations associated with mining. She believes this is a deliberate attempt to obscure the potential environmental damage and appease concerns.[15]

8.3 The High Seas

Transitioning from the focus on Exclusive Economic Zones to the high seas highlights the interconnected nature of Oceania and its cultural and ecological significance. The Clarion-Clipperton Zone (CCZ), situated beyond national jurisdictions, exemplifies the high seas as a shared space of immense opportunity and vulnerability. Unlike DSM activities within EEZs, which often centre on state-driven economic benefits, the high seas necessitate a broader, global approach—one that integrates ecological preservation, equitable resource sharing, and a recognition of the cultural and spiritual connections that Pacific communities maintain with the ocean. This perspective underscores the need for collaborative governance frameworks that honour the interconnectedness of Oceania while safeguarding its future.

The Clarion-Clipperton Zone, located between Hawaii and Mexico and bounded by the Clarion and Clipperton Fracture Zones, epitomises the tension between global economic ambitions and ecological and cultural preservation (Sharma, 2017). Since its discovery in 1954, the CCZ has become a focal point for DSM, with nations like Nauru, Tonga, and Kiribati partnering with contractors such as Nauru Ocean Resources Inc. (NORI), Tonga Offshore Mining Limited (TOML), and Marawa Research and Exploration Ltd. These companies are all subsidiaries of a global north seabed mining company, The Metals Company. Other exploration contracts in the CCZ involve countries like Belgium, China, France, Germany, Japan, Korea, the United Kingdom, and Singapore, alongside the InterOceanMetal Joint Organization, representing Bulgaria, Cuba, Czech Republic, Poland, Russia, and Slovakia. These activities are overseen by the International Seabed Authority (ISA), tasked with balancing the competing demands of economic development, environmental protection, and the cultural and spiritual values tied to the ocean (see Chap. 4, this vol.).

Through the lens of Epeli Hauʻofa's philosophy of the 'Sea of Islands' (1994), Oceania emerges not as fragmented territories but as a vast, interconnected region unified by its peoples, cultures, and histories. Hauʻofa's vision challenges the colonial legacy of division and reductionism, emphasising instead the fluidity and

[15] See also, Hosking (2024). This letter critiques the lack of transparency and community participation in seabed mining decision-making processes in the Cook Islands, highlighting the concentration of power in the hands of a few and questioning the Seabed Minerals Authority's claim of a 'social licence' to mine. Concerns include insufficient consultation with traditional leadership, the Church, and the public, environmental risks to biodiversity, and the prioritisation of mining legislation over other environmental issues. The letter underscores the need for genuine consultation and transparency.

connectivity of the Pacific. In this context, the high seas, including areas like the CCZ, are not isolated expanses but integral parts of the shared heritage and identity of Pacific Island communities. They are spaces where cultural narratives, traditional knowledge, and ecological stewardship converge, offering a counterpoint to extractive models that prioritize short-term profit over long-term sustainability.

As DSM extends its reach from EEZs to the high seas, it threatens to sever the intricate cultural and ecological relationships that Pacific peoples have nurtured for generations. The 'Sea of Islands' philosophy challenges reductionist views of the Pacific, portraying it not as fragmented territories ripe for exploitation but as an interconnected, vibrant expanse central to Oceania's identity, livelihoods, and spiritual well-being. This perspective underscores the ocean's role as a shared heritage that transcends jurisdictional boundaries and demands a holistic approach to stewardship.

The push for mining in the CCZ reflects an enduring colonial mindset that prioritises economic gain over ecological and cultural integrity. Mining threatens to replicate the extractive practices that have left lasting scars on Pacific communities. For example, phosphate mining on Banaba Island devastated 90% of its surface, transforming it into an industrial wasteland and forcibly displacing the Indigenous Banaban people. This relocation fractured their cultural ties to their ancestral homeland, creating intergenerational trauma that continues to impact their identity and well-being. In the book *Inflamed*, Rupa Marya and Raj Patel (2022) argue that colonial extractivism is not merely a relic of the past but an ongoing systemic force that disrupts ecosystems, displaces communities, and exacerbates inequality and health disparities. Without careful governance and cultural sensitivity, the CCZ risks becoming another chapter in this exploitative legacy, reinforcing patterns of ecological destruction and cultural erasure in the pursuit of short-term economic benefit.

Beyond its mineral wealth, the CCZ is one of Earth's most biodiverse regions, with an astounding 70–90% of collected species new to science. The polymetallic nodules that attract mining interest are far more than inert resources; they serve as critical habitats for diverse life forms, ranging from microbial communities to deep-sea invertebrates. This biodiversity is deeply interwoven with the cultural and spiritual worldviews of Pacific communities, who see the ocean as a living, dynamic entity sustaining both ecological and human well-being. Without deliberate governance measures, the CCZ risks becoming another chapter in the legacy of extractive exploitation, undermining both ecological resilience and cultural heritage.

Tongan scholar Tēvita O Ka'ili's discussion of *vä*, or 'sacred relationality', provides a profound lens for understanding the interconnectedness of people and their environment (Ka'ili, 2005, 2017).[16] *Vä* underscores the inseparability of cultural

[16] Tēvita O Ka'ili has extensively explored the concept of vā, or 'sacred relationality', in his scholarly work. In his 2005 article, Ka'ili examines how Tongans maintain and nurture social relationships and spaces, emphasizing the importance of vā in Tongan culture. Additionally, in his 2017 book, *Marking Indigeneity: The Tongan Art of Sociospatial Relations*, Ka'ili delves into the Tongan art of sociospatial relations, further elucidating the significance of vā in the context of Tongan identity and community.

identity and ecological stewardship, offering a framework that highlights the need for governance models that respect the cultural and ecological dimensions of the high seas. This relational perspective calls for global policies that balance innovation with the preservation of cultural and ecological legacies, ensuring that the ocean remains a shared space of connectivity and life for future generations. Similarly, the Māori principle of *whakapapa* links all things through ancestral lineage, establishing responsibilities to both people and the natural world. From these relationships arises *kaitiakitanga*—a duty to protect and manage ancestral domains for future generations.

Oceania's leadership in community-based protected areas exemplifies the integration of these values into modern conservation initiatives. These efforts prioritise sustainable use and conservation, drawing on cultural kinship, customary governance, and ecological stewardship to achieve a balance between development and preservation. In Aotearoa, *kaitiakitanga* embodies this balance, intricately connected to *rangatiratanga*, the right to self-determination outlined in Te Tiriti o Waitangi (Jackson et al., 2017). Together, these Indigenous frameworks offer powerful models for high seas governance, demonstrating that economic innovation can and must coexist with the preservation of cultural and ecological legacies, intertwined with the cultural and spiritual perspectives of many Pacific communities, who view the ocean as more than a resource—it is a living, dynamic entity that sustains both ecological and human well-being (McAllister et al., 2023).

These risks compel us to adopt a broader perspective—one that aligns scientific research with cultural knowledge and wisdom. The CCZ is not just a repository of minerals; it is part of a larger, interconnected oceanic system imbued with cultural narratives and ecological significance. Pacific philosophies remind us that the ocean is a shared space of connection and responsibility. As stated, Pacific Islanders have long mapped sea routes, named places across vast distances, and cultivated spiritual relationships with the ocean. Far from being resource-poor, Oceania's wealth lies in its ecological and cultural knowledge systems, which reflect a profound understanding of the natural world and its interconnectedness with human life. Integrating these cultural perspectives with precautionary scientific approaches offers a pathway to protect the CCZ's fragile ecosystems while honouring the deep relationship between Pacific communities and their ocean. In this way, we can envision a future where both biodiversity and cultural integrity are safeguarded for generations to come.

Colonial and industrial narratives, however, have overshadowed these perspectives. The CCZ, like much of Oceania, has been framed as a resource-rich frontier ripe for exploitation, perpetuating the history of environmental extraction that devastated Indigenous ways of life. For instance, the loss of Indigenous languages—resulting from colonial assimilation policies—has severed critical connections between people and their environments. These languages often encoded sophisticated ecological knowledge, contributing to the erosion of cultural identity and well-being.

Resistance to DSM is deeply rooted in Oceanian cultural identity and place-attachment. The Kumulipo, a Hawaiian creation chant, exemplifies this relationship. It recounts the origins of the world, emphasising the interconnectedness of humans and the ocean. Uncle Sol uses the Kumulipo to advocate for the protection of the deep sea, asserting that its genealogical ties to Hawaiians make it a sacred site, not a vacant expanse for exploitation. This perspective aligns with Indigenous knowledge systems that emphasise holistic approaches to health and well-being. *Inflamed* contrasts this with Western medicine's reductionist approach, which isolates the individual body from its ecological and social context. Indigenous frameworks, in contrast, reveal the intricate web of relationships connecting human health with environmental health.

8.4 Conclusion

The transformation of marine ecosystems through industrial activities like mining and shipping poses huge risks in the Pacific region.

The disruption of environmental and food connectivity, as seen in the links between land-based pollution and ciguatera fish poisoning (CFP) in Fiji and the Cook Islands, resonates with broader challenges posed by sediment plumes generated by DSM operations. Just as terrestrial contaminants from mining activities and agricultural runoff can destabilise marine ecosystems by promoting harmful algal blooms, DSM sediment plumes extend these risks to previously undisturbed deep-sea environments, upon which communities rely for their livelihoods and identity. The scale of these plumes—discharging tens of thousands of cubic meters of sediment daily, as in the Clarion-Clipperton Zone—threatens migratory species and regional food chains, critical to both ecological integrity and the wellbeing, identity and subsistence needs and practices of Pacific communities.

But what are some solutions and examples of management? Oceania leads the world in community-based protected areas, which account for a large portion of all protected sites in the region. These initiatives prioritise sustainable use alongside conservation goals, rooted in cultural kinship, customary governance and ecological stewardship. They offer a model for balancing development with the preservation of cultural and ecological integrity. Similarly, Vanuatu demonstrates how integrating traditional knowledge and values into national policies strengthens ocean stewardship. Vanuatu's Marine Spatial Plan (MSP) serves as a cornerstone of Vanuatu's inclusive approach, incorporating zoning systems that empower local communities through Community Conservation Areas and no-take zones, while protecting biodiversity and supporting sustainable resource use. Traditional knowledge, documented through initiatives like the Vanuatu National Traditional Knowledge Indicators Booklet, plays a pivotal role in guiding climate adaptation and resource management. By embedding Indigenous practices such as weather forecasting based on environmental cues into contemporary frameworks like the MSP, Vanuatu demonstrates how Indigenous knowledge can enhance resilience and ecological stewardship.

Addressing the multitude of challenges requires listening to, respecting, and integrating cultural approaches and traditional ecological knowledge into ocean governance frameworks. To conclude, we invite readers to reflect on the words of the Pacific Theological College Principal, Rev. Professor Upolu Vaai and his call on all churches, all Christians, all Pacific Islanders to stand together to protect the ocean:

> We don't have the ocean, we are ocean. We don't have the land, we are land. Watery to our very core. Earthy to our very being. The Moana is symbolic of the Spirit. Dynamic. Connected. Alive. Yet Groaning. Constantly seeking our hand of compassion in the age of destruction. The moana groans for restoration. Yet the Spirit comes to rouse the people. The Spirit births new voices. She sparks movements of resistance. She creates currents of contradictions to exploitation. She disturbs our complacency. She inspires new moana consciousness. She makes us passionate people for the one groaning. With a wild and unruly practice of hope. Lord God, reconnect us to the moana. Show us a new Alaimoana. A new pathway. Guided by stars of hope that guided our forebears. Create in us a new consciousness of care and connectedness guided by your love. Reconnect us to our watery home. Our fluid identity. In the name of the God who liberates us from the gods of money. Amen! (Raj, n.d.)

References

Arato, J., Jaeckel, A., & Lixinski, L. (2024). On the cultural stakes of deep seabed mining. *AJIL Unbound, 118*, 67–71. https://doi.org/10.1017/aju.2024.9

Boucher, D. L., Macumber, B., & James-Proud, M. (2024, October 25). Kiribati fishermen tell how they survived being lost at sea for 33 days and solved a family mystery. *ABC Pacific*. https://www.abc.net.au/pacific/uein-buranibwe-was-lost-at-sea-for-33-days-story-of-survival/104491232.

Childs, J. (2019, February 19). Deep sea mining threatens indigenous culture in Papua New Guinea. *The Conversation*. http://theconversation.com/deep-sea-mining-threatens-indigenous-culture-in-papua-new-guinea-112012.

Cook Islands News. (2023, November 22). Sovereign rights not enough, regional cooperation needed for deep-sea mining: TIS. *Islands Business*. https://islandsbusiness.com/news-break/deep-sea-mining-3/.

DOSI. (2021). *The necessity of traditional knowledge for management of deep-seabed mining* [Policy Brief]. https://www.dosi-project.org/wp-content/uploads/072-DOSI-Policy-brief-Traditional-Knowledge-for-Management-of-DSM-long-V11.pdf.

Drazen, J. C., Smith, C. R., Gjerde, K. M., Haddock, S. H. D., Carter, G. S., Choy, C. A., Clark, M. R., Dutrieux, P., Goetze, E., Hauton, C., Hatta, M., Koslow, J. A., Leitner, A. B., Pacini, A., Perelman, J. N., Peacock, T., Sutton, T. T., Watling, L., & Yamamoto, H. (2020). Midwater ecosystems must be considered when evaluating environmental risks of deep-sea mining. *Proceedings of the National Academy of Sciences, 117*(30), 17455–17460. https://doi.org/10.1073/pnas.2011914117

Emberson-Bain, ʾA., Bain-Vete, M., Bain-Vete, S., & Bain-Vete, A. (2023). *Cries from the Moana*. Hakautapu Publishing.

Fiji One News Team. (2024, August 9). International acclaim for "Cries from Moana." *Fiji News*. https://fijionenews.com.fj/international-acclaim-for-cries-from-moana/.

Fuatai, T. (2024, November 23). Deep-sea mining's charm offensive in the Cook Islands. *E-Tangata*. https://e-tangata.co.nz/comment-and-analysis/deep-sea-minings-charm-offensive-in-the-cook-islands/.

Hau'ofa, E. (1994). Our Sea of Islands. *The Contemporary Pacific, 6*(1), 148–161. https://www.jstor.org/stable/23701593

Hikuroa, D. (2020). Matauranga Maori and its role in coastal management. In C. Hendtlass, S. Morgan, & D. Neale (Eds.), *Coastal systems and sea level rise: What we look for in the future*. New Zealand Coastal Society. https://www.coastalsociety.org.nz/assets/Publications/Special-Issues/SP4-Low-res-version.pdf

Hosking, J. (2024, October 11). Letter: Social licence for deep sea mining. *Cook Islands News*. https://www.cookislandsnews.com/internal/opinion/letters-to-the-editor/letter-social-licence-for-deep-sea-mining/.

Ingersoll, K. E. (2016). *Waves of knowing: A seascape epistemology*. Duke University Press.

Jackson, A. M., Mita, N., & Hakopa, H. (2017). *Hui-te-ana-nui: Understanding kaitiakitanga in our marine environment* (Sustainable Seas National Science Challenge Project). https://www.sustainableseaschallenge.co.nz/assets/dms/Reports/Hui-te-ana-nui-Understanding-kaitiakitanga-in-our-marine-environment/HTAN-report-2017.pdf.

Ka'ili, T. O. (2005). Tauhi vā: Nurturing Tongan Sociospatial Ties in Maui and Beyond. *The Contemporary Pacific, 17*(1), 83–114. https://www.jstor.org/stable/23721933

Ka'ili, T. O. (2017). *Marking indigeneity: The Tongan art of sociospatial relations. University of Arizona Press*. https://doi.org/10.2307/j.ctt1t89kr9

Ka'ili, T. (2024, February). *Fonua: Intersecting Cosmogony and Ecology* [2024 David O. Mckay Lecture]. https://speeches.byuh.edu/david-o-mckay-lectures/intersection-cosmogony-and-ecology.

Kessler, R.,2024. Opinion: Communities worry anew as PNG revives seabed mining plans. In Islands Business, March 21, 2024.

Lako, J. V., Naisilisili, S., Vuki, V. C., Kuridrani, N., & Agyei, D. (2023). Local and traditional ecological knowledge of fish poisoning in Fiji. *Toxins, 15*(3), 223. https://doi.org/10.3390/toxins15030223

Lala's LALINOA In-Depth Conversation with Tevita Ka'ili: Exploring Ancient Tongan Beliefs Before Christianity and Their Influence on Traditional Practices Today. (2024, September). [Broadcast]. Radio Nukuálofa 88.6FM. https://www.youtube.com/watch?v=Z-V3MfhRBRc.

Marx, W., & Kiernan, E. (2024a, September 19). Papua New Guinea leaders struggle to monitor deep-sea mining activities off its coast. [Broadcast]. In *News hour*. PBS. https://www.pbs.org/newshour/show/papua-new-guinea-leaders-struggle-to-monitor-deep-sea-mining-activities-off-its-coast.

Marx, W., & Kiernan, E. (2024b, September 18). Papua New Guinea locals concerned over deep-sea mining's impact on culture and environment. [Broadcast]. In *News hour*. PBS. https://www.pbs.org/newshour/show/papua-new-guinea-locals-concerned-over-deep-sea-minings-impact-on-culture-and-environment.

Marya, R., & Patel, R. (2022). *Inflamed: Deep medicine and the anatomy of injustice* (First paperback ed.). Picador.

McAllister, T., Hikuroa, D., & Macinnis-Ng, C. (2023). Connecting science to indigenous knowledge: Kaitiakitanga, conservation, and resource management. *New Zealand Journal of Ecology*. https://doi.org/10.20417/nzjecol.47.3521

Nichols, P. (1992). *Sharks* (No. 92/66). FFA (Pacific Islands Forum Fisheries Agency). https://spccfpstore1.blob.core.windows.net/digitallibrary-docs/files/69/690fca052ecccef8edf66bbf2e5fdfd6.pdf?sv=2015-12-11&sr=b&sig=e44UALp00Jao%2BDtPZo6uY7UmstP6mhT2dlNlSjyxkOY%3D&se=2025-06-05T20%3A40%3A58Z&sp=r&rscc=public%2C%20max-age%3D864000%2C%20max-stale%3D86400&rsct=application%2Fpdf&rscd=inline%3B%20filename%3D%22FFA_1992_066.pdf%22.

Philips, H. (2019). *A sea of voices: Deep sea mining and the Solwara 1 Project in Papua New Guinea* [MA Thesis, University of Waikato]. https://researchcommons.waikato.ac.nz/entities/publication/e60d590f-82d4-4e88-ab31-5f703e4d1be0.

Pili Kahoʻohalahala, S. (2024, December 16). *What global deep sea mining negotiations are missing*. Greenpeace International. https://www.greenpeace.org/international/story/66154/uncle-sol-pacific-voices-say-stop-deep-sea-mining/.

Raj, B. (n.d.). *A CALL FOR THE OCEAN – Pacific Theological College*. Retrieved December 16, 2024, from https://ptc.ac.fj/a-call-for-the-ocean/.

Rongo, T., Bush, M., & Van Woesik, R. (2009). Did ciguatera prompt the late Holocene Polynesian voyages of discovery? *Journal of Biogeography, 36*(8), 1423–1432. https://doi.org/10.1111/j.1365-2699.2009.02139.x

Segalov, M. (2023, May 21). Is it too late to halt deep-sea mining? Meet the activists trying to save the seabed. *The Observer*. https://www.theguardian.com/environment/2023/may/21/is-it-too-late-to-halt-deep-sea-mining-the-activists-trying-to-save-the-seabed.

Sharma, R. (Ed.). (2017). *Deep-sea mining: Resource potential, technical and environmental considerations*. Springer International. https://doi.org/10.1007/978-3-319-52557-0

Teaiwa, K. M. (2014). *Consuming Ocean Island: Stories of people and phosphate from Banaba*. Indiana University Press.

Torrente, F., Bambridge, T., Planes, S., Guiart, J., & Clua, E. G. (2018). Sea swallowers and land devourers: Can shark lore facilitate conservation? *Human Ecology, 46*(5), 717–726. https://doi.org/10.1007/s10745-018-0018-7

Tuari'I, M., & Worth, H. (2024). *Opinion poll on deep-seabed mining in the cook islands, Analysis Report 2024*.

Chapter 9
Negotiating Human Cultural Relations with the Sea in South Africa and Namibia

Rosabelle Boswell

Abstract Along the African coast are a diversity of peoples, many of whom hold a unique cultural relationship with the sea. In the last few decades, Environmental and Social Impact Assessments (ESIA), a legal requirement in many African countries, are, in some instances, now including Cultural Heritage Impact Assessments (CHIA). In the following a discussion on the value of CHIA to ESIA is offered, as well as a discussion on human cultural relations with the sea and deep ocean in Africa. The discussion is anchored in social science theories of posthumanism (Braidotti, 2016) and pluriversal knowledges (Escobar, 2018), theoretical strands which posit that anthropocentric (human-centred) perspectives prioritise humanistic frames of world and obscure the agency of nature and inanimate things. Considering transmaterial and pluriversal knowledges, humans may be able to perceive and treat the ocean not as an asset but as a living entity that has intersubjective communication with human kind. To contextualise the discussion, two case studies are offered, one of spiritual veneration and the place of the seabed as the final resting place of ancestors in South Africa and a second case study of severed cultural relations with the sea as a spiritual entity in the aftermath of colonial rule in Namibia. The chapter begins with a brief overview of the history and anthropology of the two geographical sites and continues with a presentation of the case studies. Finally, the chapter reviews the place of ESIA and CHIA in proposed offshore development in these two countries.

Keywords Intangible cultural heritage · Namibia · South Africa · Environmental and social impact assessments · Cultural heritage impact assessment

R. Boswell (✉)
Coastal and Marine Research Institute, Ocean Sciences Campus, Nelson Mandela University NRF Grant UID 129962, Gqeberha, South Africa
e-mail: rose.boswell@mandela.ac.za

C. Jarvis (ed.), *Threats to Our Ocean Heritage: Deep Sea Mining*, SpringerBriefs in Archaeology, https://doi.org/10.1007/978-3-031-98238-5_9

Africa is a vast continent with a coastline of 18,950 miles. Along the African coast are a diversity of peoples, many of whom hold a unique cultural relationship with the sea. In the last few decades, Environmental and Social Impact Assessments (ESIA), a legal requirement in many African countries, are, in some instances, now including Cultural Heritage Impact Assessments (CHIA). In the following, a discussion on the value of CHIA to ESIA is offered, as well as a discussion on human cultural relations with the sea and deep ocean in Africa. The discussion is anchored in social science theories of posthumanism (Braidotti, 2016) and pluriversal knowledges (Escobar, 2018), theoretical strands which posit that anthropocentric (human-centred) perspectives prioritise humanistic frames of world and obscure the agency of nature and inanimate things. Considering transmaterial and pluriversal knowledges, humans may be able to perceive and treat the ocean not as an asset but as a living entity that has transmaterial connections with humanity. To contextualise the discussion, two case studies are offered, one of spiritual veneration and the place of the seabed as the final resting place of ancestors in South Africa and a second case study of severed cultural relations with the sea as a spiritual entity in the aftermath of colonial rule in Namibia. The chapter begins with a brief overview of the history and anthropology of the two geographical sites and continues with a presentation of the case studies. Finally, the chapter reviews the place of ESIA and CHIA in proposed offshore development in these two countries and offers something to consider with deep sea mining.

9.1 History and Anthropology in Namibia and South Africa

The indigenous peoples of southern Africa, known now as the Khoisan, are considered to be the descendants of the earliest modern human beings, *Homo Sapiens*, to have lived on Earth. In this regard there are important archaeological sites in South Africa (including coastal) that are now designated as the 'cradle of humankind'. For instance, in July 2024, the South African government successfully nominated the Pinnacle Point Site Complex (which includes associated caves of archaeological significance) as a UNESCO World Heritage Site. The site is now one of more than 1000 UNESCO World Heritage Sites which are recognised as being universally valuable and expressive of global human creativity. Pinnacle Point Site Complex (PPSC), situated along the southern Cape coast of South Africa, is indicative of 160,000 years of human-nature exchange and thus, according to the palaeoanthropologist Curtis Marean, is the site of the earliest aquatic hunter-gatherers and practitioners of coastal cultural heritage (Marean, 2022). Today, the Khoisan peoples consist of many subgroupings spread across southern Africa. They have been extensively studied and much is now known about their territorial creativity and spirituality.

The Khoisan peoples have also long traversed the borders between Namibia and South Africa. Today, there are shared languages, food heritages, musical traditions and beliefs between indigenous Khoi and San groupings in Namibia and South Africa. Anthropological and historical studies of the two countries reveal the extent to which the First Nations peoples were socially and economically marginalised during colonial rule. A critical overview of the position of the Khoisan peoples after colonialism and the importance of including them and their perspectives on ocean management has been offered elsewhere (Boswell & Thornton, 2021).

During the colonial period, ethnocentric (discriminatory) accounts of the Khoisan (and other indigenous groupings), portrayed them as subhuman (Gould, 1981; Gordon & Sholto-Douglas, 2000), positioning the Khoisan peoples at the base of rigid, of a racialised hierarchy. During apartheid in South Africa (1948–1994), the government also construed Khoisan peoples' knowledge of nature integrated perspective on as evidence of atavism. To this end, the South African government used some Khoisan men as trackers of 'enemy' forces during the war between apartheid South Africa and Namibia's liberation fighters (Gordon & Sholto-Douglas, 2000). Since Namibia was also part of the apartheid state and was designated Southwest Africa, it was also subjected to racial segregation. Indigenous peoples such as the Ovambo and Nama were forcibly relocated from the coast to inland towns such as Tsumeb and Tsintsabis, or desert locations such as Gobabeb, near Swakopmund. Ethnographic research (2024) for a Cultural Heritage Impact Assessment (CHIA) in Luderitz, Namibia also revealed the impact of genocide on Namibia's indigenes and what communities desire as they pursue long-term social justice. Presently, Non-Governmental Organisations (NGOs) such as Sister Namibia, situated in Windhoek, document the historical and present legacy of Khoisan marginalisation and discrimination. Sister Namibia documents how Khoisan women were often the ones experiencing the most discrimination and violation as a result of the racist and sexist practices of segregationist regimes in Namibia and South Africa.

Thus, and as this Chapter shows, South Africa and Namibia are linked by a long history of colonisation, cultural exchange, and economic trade. They also share many current social features. Both countries attained independence from colonial rule in the 1990s, both opposed racialised and segregationist political regimes that led to the impoverishment and the social marginalisation of the mostly indigenous peoples. Both experienced a violent, negotiated transition to democracy. Pre-1990s accounts of indigenous peoples (specifically the Khoisan) present them as mainly inland peoples. Although historically and ethnographically rich, these accounts have tended to advance the legacy of indigenous exclusion and marginalisation in political and conservation leadership. Thus (See Boswell & Thornton, 2021), it is stated that:

> Today, it is evident that indigenous knowledge forms need to be excavated, restored and included in considerations of ocean conservation, not only because of the historical exclusion of indigenous peoples but because current approaches to ocean conservation, and ultimately ocean governance, are failing. The world is facing a catastrophic climate crisis coupled with deepening inequality and exclusion.

It is well recognised that social identities and the cultural practice of identity are subject to social construction, meaning that powerful individuals within a group may dominate the allocation of social meaning, narration of the past and ultimately the crafting of intangible cultural heritage (ICH). To further complicate matters, coastal real estate and both offshore and onshore economic development in Namibia are now attracting growing investor interest. At the same time, environmental lobbies are also shaping marine and coastal agendas. As a result, the cultural identity and the valuation of the ocean are no longer solely determined by local communities but are now subject to competing economic and environmental interests. Local cultural identities remain dynamic. Rituals, beliefs and symbolisms continue to manifest in public demonstration and commemoration of identity and history. Ethnographic studies in the last decade or so (Verbuyst, 2022; Gabie, 2014), tell that while indigenous communities present an integrated relationship with nature and the sea, these relationships are subject to change. Change becomes apparent for example when a community is forced (because of apartheid law for example) to leave a coastal area and to abandon spiritual and ritual connection with the sea.

A final but equally important factor in the narrative of identity, culture and ocean management in Africa is the sheer diversity of indigenous groupings. In this discussion, more attention is given to the Khoisan peoples in both Namibia and South Africa; a grouping that is by no means unified or homogeneous. But beyond this group, there are some 60 million South Africans, many of whom live at the coast or have migrated to major cities there since the fall of apartheid. The Nguni descendent Xhosa peoples of South Africa for example, are often publicly presented as inshore communities focused on land and territorial species. However, and as some ethnographic studies (Giampicolli & Kalis, 2012; Boswell, 2022) show, the Xhosa (and other Nguni descended groups) have long cultivated a rich relation with marine animals, oceans and rivers (Bernard, 2010). Some of these aspects are discussed next.

9.2 Coastal Ethnography

From 2022 to 2024, the author and their research team conducted anthropological research in coastal Namibia and South Africa, producing more than 500 interviews on human cultural relations with the sea and coast (see Figs. 9.1 and 9.2).

The interviews in South Africa reveal a rich human relationship with the sea and coast. Those interviewed spoke of the physics of the sea (Boswell, 2022) and its contribution to human buoyancy in contrast to the 'nothingness' of air. They spoke of the rhythmic sounds of waves as a regulatory counterpoint to human-made (polluting) sonic states. The ocean is, for the many people encountered beyond identifiable indigenous groupings, a watery refuge from the vicissitudes of modern life in unequal and increasingly unpredictable economies. For most, it is a source of physical and psychological wellness, a natural balm for the psychological trauma arising from the colonial and apartheid era. In brief, during the years of anthropological research, both Namibians and South Africans living at the coast spoke of the sea as a source of leisure, as culturally meaningful place and/or spiritual dimension. For the tradition revering Xhosa peoples of South Africa, the sea is believed to be the final resting place of the ancestors. It is to be visited with ritualised offerings and, more importantly, to be approached with caution, as it may seek to retain those with whom it has affinity. The traditional Xhosa believe that if twins visit the ocean for

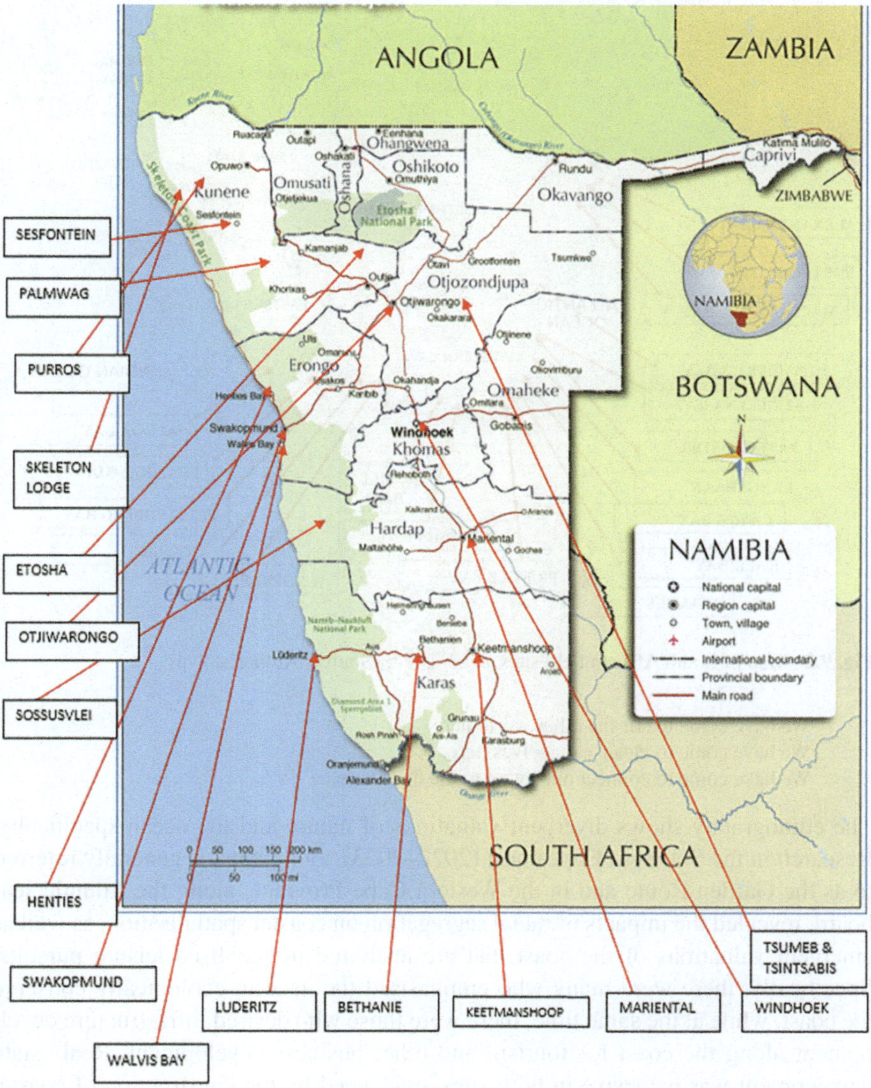

Fig. 9.1 Map of Namibia indicating field research sites from 2022 to 2024. (Source: Author's own)

example, it is important that special offerings be made as there is a likelihood that one will be 'retained' by the ocean. The spiritual dimension of the ocean is such that a national poet, Koleka Putuma immortalized this aspect in her poem, 'Water', in which she contrasts the experiences of European descendants with that of Africans, saying,

…For you, the ocean is for surfboards, boats and tans
and all the cool stuff you do under there in your bathing suits and goggles.
But we,
we have come to be baptised here.

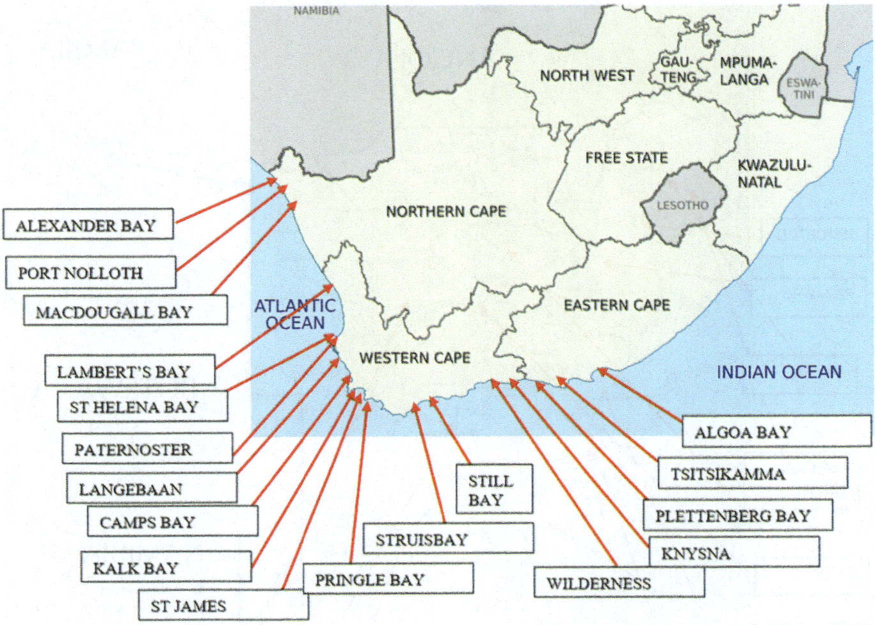

Fig. 9.2 Map of South African field sites 2022–2024. (Source: Author's own)

> We have come to stir the other world here
> We have come to cleanse ourselves here.
> We have come to connect our living to the dead here.

The ethnography shows divergent valuations of nature and the ocean specifically. Research in the Southern Cape coast (2022–2023), along what is generally referred to as the Garden Route and in the Western Cape Province, along the Atlantic seaboard, revealed the impacts of racial segregation on coastal spatialisation, as well as emergent valuations of the coast that are anchored in ritualised leisure pursuits. Specifically, there were many who emphasised the need to ecologically conserve the coast, while at the same time, there were those who desired infrastructure development along the coast for tourism and other business development. Real estate development was extensive in both sites, evidenced by the construction of coastal luxury estates, residences and eco-estates. These developments ultimately secure the coastline for leisure that is aligned with the social needs of more wealthy groups. However, and as analysed, Chapter 2 of Constitution of South Africa seeks to address apartheid-era inequalities and promote human flourishing. Many of the inequalities have yet to be fully addressed.

The reason for noting the above is as follows: Presently, South Africa's oceans appear to be a contested landscape. On the one hand, one finds those who are wealthier, have greater access to the coast (i.e., those of European descent) and possibly the 'luxury' of considering coastal and ocean conservation. On the other hand, are the many applications for offshore exploration and development, which seek to leverage the oceans for resource use. Third, there are environmental NGOs, focused on mitigating the impacts of global climate change. In this contested coastal landscape, in South Africa, a 3000 km coastline predominantly inhabited by those privileged during apartheid, robust response to nascent offshore oil and gas projects and insufficient plans to address long-term inequalities at the coast (and in the economy), lead to continued poverty for millions of South Africans.

The research findings in Namibia present a different scenario. Namibia is country where there is significant inland mining (for precious minerals and metals). Much of the land is also allocated to nature conservancies. The population is merely 2.5 million spread across a vast territory. With an extremely low population density and fewer coastal sites of economic importance, the Namibia government appears to experience less opposition to proposed offshore exploration and development. Moreover, unlike South Africa, indigenous Namibians (the San peoples) perceive spiritual and cultural connections with inland phenomena, such as subterranean features (i.e., the *Wondergat*) and the wind. Ethnographic research in northeastern Namibia in 2023, revealed that forced relocations from the coast to inland towns and settlements during the period of apartheid under the Odendaal Commission of 1964 meant that many indigenes were severed from a cultural existence with the coast.

In late 2023, the research team invited some San descendants to undertake a 'return to the sea', as interviews in one Namibian village revealed rich stories of rituals and beliefs regarding the ocean. Specifically, one chief recounted how his grandfather spoke of the importance of greeting the sea when arriving at the shore for the first time. He said that once one greets the sea and asks for protection and permission, the sea will protect and allow the individual to proceed with his or her planned endeavours. The exact words of the San descendant were, 'the ocean has immense power, she is generous but we must only ask'. Interestingly, the same sentiment was echoed by a coastal dweller in South Africa. She said, that 'we need to ask the ocean for permission to use what she has'. Further interviews with other chiefs and elders in both countries revealed the importance of conversation and ritual engagement with nature. The data revealed what social scientists now say is a transmaterial and posthuman (Braidotti, 2016) relationship with nature; one that recognises the intersubjectivities of the living and 'non-living' worlds. What is also interesting is that the transmaterial elements of human connection to the ocean is now manifesting in groupings that are not indigenous and do not routinely practice a ritual relationship with the sea. During fieldwork in South Africa, for example, it was found that women are developing both relational and ritualised relations with the ocean. Specifically, in the Western Cape Province, there were women who, during the full moon, would wade into the sea to reconstitute their feminine essence and strength and women who anthropomorphised the sea, perceiving feminine qualities in it. In this regard, some women said that the reason why women cannot be at sea is because the sea is a woman. She prefers men and becomes jealous when women are also at sea. Others expressed the view that as a woman, the sea reflects the imagined temperament of women, which they classified as temperamental, soft and accommodating but also potentially violent and rejecting.

Of value and interest to the discussion is that the author has also engaged with offshore operators via Environmental and Social Impact Assessments (ESIA). ESIA have, for the last ten years at least, formed part of the Environmental Impact Assessment (EIA) processes of South Africa and Namibia. In these two countries, ESIA form part of the democratic process of inclusion in ocean and nature management processes. Briefly, and until recently, ESIA in both countries have not included a Cultural Heritage Impact Assessment (CHIA), that considers Intangible Cultural Heritage in the coastal context. ICH are those elements of human culture and cultural practice that are not palpable. They consist of rituals, beliefs, symbolism, and expressions that are of enduring cultural value and may, in due course, be recognised by UNESCO and nominated for UNESCO World Heritage Status. In conducting CHIA, my research team and I seriously considered the presence and role of ICH in shaping human relations with the sea in Namibia and South Africa.

In 2022, Totalenergies requested CHIA for several of its proposed offshore block exploration projects. The request triggered similar calls for CHIA by other developers. Of relevance to this discussion is that in each CHIA report, there are key mitigation measures for the recognition of cultural relationships with the sea, as well as a recommendation that the necessary conditions/indigenous forums be supported to discern which cultural practice or rituals will be necessary for engagement with the ocean. This mitigation measure has been seriously considered and in most instances, accepted by operators for whom a CHIA has been prepared between 2022 and 2024. This is significant because it indicates formal, offshore *industry* recognition of (1) the right to cultural diversity (2) the importance of culture to the protection of human rights and (3) the existence of human cultural relations with the sea and coast. Anthropologically, recognition of a cultural relationship does not mean that any subsequent cultural intervention will suffice, or achieve the desired cultural aim, which in this case means receiving the spiritual and cultural support to implement the project. What the case studies and the CHIA reports do demonstrate is that ocean governance and management are negotiated processes in highly contested and dynamic landscapes. Both ocean governance and management involve many stakeholders and multiple valuations of the ocean. The CHIA reports and the ethnographic research informing them, also show that the voices of indigenous communities are valuable and are becoming more audible in the still unequal societies of South Africa and Namibia.

9.3 Conclusion

The Namibia and South Africa case studies suggest that it is important to consider the significance of human cultural relations with the ocean and other bodies of water, and specifically, how proposed exploration and development may affect cultural relations with the ocean. The ethnographic data also shows that humans have a very rich relationship with nature, humans spiritually and culturally connect with diverse bodies of water, such as the shore and rivers. They develop rituals that includes these natural elements and retain memory of historical practices involving nature. Regarding proposed and future plans to extract resources from the ocean and the seabed, there is need to consider not only these ritual relations but the unequal and contested socioeconomic landscapes in which cultural relationships with the sea and coast, exist. Commercial uses of the ocean (i.e., offshore oil and gas projects) are not the only entities impacting coastal communities. There is a long legacy of coastal dispossession and inequality that continues to affect indigenes and others impoverished by colonial and apartheid rule. Considering the ethnography directly, one learns that human cultural relations with the sea evoke diverse logics. These include acceptance of ocean 'sentience' and acceptance of the ocean as a spiritual domain. This means that, at the very least, those seeking to explore and develop the ocean for economic use, need to include a diversity of stakeholders in the consultation process. As the case studies presented here show, however, it is not easy to navigate a socially and historically complex set of stakeholders. In the case of South Africa, there are vested (economic) interests in maintaining the status quo in still segregated coastal settings. Unless robustly engaged, the coastal status quo is unlikely to change or become more inclusive of indigenous peoples and their world views. And yet, the ethnographic research in Namibia and South Africa are showing that, there is a rich and diverse epistemology of the ocean in southern Africa. These knowledge forms are vital not only to a deeper understanding of cultural diversity in Africa but also to the development of future philosophies of nature which, if mainstreamed, can advance a more democratic and sustainable ocean care for future generations.

References

Bernard, P. S. (2010). *Messages from the deep: Water divinities, dreams and diviners in Southern Africa*. Unpublished PhD Thesis, Rhodes University, South Africa.

Boswell, R. (2022). Salted identities: Biocultural heritage for a rehumanized ocean management in South Africa. *Anthropology and Humanism, 47*(2), 363–380.

Boswell, R., & Thornton, J. L. (2021). Including the Khoisan for a more inclusive blue economy in South Africa. *Journal of the Indian Ocean Region, 17*(2), 141–160.

Braidotti, R. (2016). Posthuman critical theory. In D. Banerji & M. R. Paranjape (Eds.), *Critical posthumanism and planetary futures* (pp. 13–33). Springer.

Escobar, A. (2018). *Radical interdependence, autonomy, and the making of worlds*. Duke University Press.

Gabie, S. (2014). *Khoisan ancestry and coloured identity: A study of the Korana Royal House under Chief Josiah Kats*. Unpublished Thesis, University of the Witwatersrand.

Giampicolli, A., & Kalis, J. (2012). Community-based tourism and local culture: The case of the amaMpondo. *PASOS Revista de Turismo y Patrimonio Cultural, 10*(1), 173–188.

Gordon, R., & Sholto-Douglas, S. (2000). *The Bushman myth: The making of a Namibian underclass*. Avalon Publishing.
Gould, S. (1981). *The mismeasure of man*. W. W. Norton.
Marean, C. (2022). The roots of our evolutionary blue heritage. In R. Boswell, D. O'Kane, & J. Hills (Eds.), *Blue heritage: Global perspectives on ocean histories and cultures*. Palgrave Macmillan.
Verbuyst, R. (2022). *Khoisan consciousness: An ethnography of emic histories and indigenous revivalism in post-apartheid Cape Town*. Brill.

Chapter 10
Conclusions and Steps Forward

Charlotte Jarvis and Bobbi-Jo Dobush

Abstract The case studies in this book and the interventions from the International Seabed Authority (ISA) (see Appendix) meetings showcase the types of Ocean Heritage, natural and cultural, tangible and intangible, that may be lost if deep sea mining (DSM) goes ahead. The chapters also provide some suggestions to protect heritage in the deep sea for future generations. For example, exclusion zones, as seen with dredging regulations (Chap. 5, this vol.). What is needed most, is surveys of the Ocean Heritage in the deep, especially in zones proposed for mining. We cannot protect heritage we do not know about. These surveys must be done to standards that archaeologists can work with, and the results should be publicly available. Underwater cultural heritage (UCH) must be included in these surveys along with natural heritage and environmental impact assessments. This is because UCH can support ecological marine biodiversity and help boost sea connectivity. Understanding UCH, including intangible UCH, is also key to respecting international commitments to human rights (including intergenerational rights, the right of public participation, the right to culture, and Indigenous peoples' rights).

Keywords Underwater cultural heritage · Deep sea mining · Deep seabed mining · Deep-sea mining, Intangible cultural heritage · International seabed authority · Underwater archaeology

C. Jarvis (✉)
The Ocean Foundation, Washington, DC, USA

B.-J. Dobush
Salt Horizon, LLC, San Diego, CA, USA

C. Jarvis (ed.), *Threats to Our Ocean Heritage: Deep Sea Mining*,
SpringerBriefs in Archaeology, https://doi.org/10.1007/978-3-031-98238-5_10

The case studies in this book and the interventions from the International Seabed Authority (ISA) (see Appendix) meetings showcase the types of Ocean Heritage, natural and cultural, tangible and intangible, that may be lost if deep sea mining (DSM) goes ahead. The chapters also provide some suggestions to protect heritage in the deep sea for future generations. For example, exclusion zones, as seen with dredging regulations (Chap. 5, this vol.). What is needed most, is surveys of the heritage in the deep, especially in zones proposed for mining. We cannot protect heritage we do not know about. These surveys must be done to standards that archaeologists can work with, and the results should be publicly available. Underwater cultural heritage (UCH) must be included in these surveys along with natural heritage and environmental impact assessments. This is because UCH can support ecological marine biodiversity and help boost sea connectivity. Kirstin Meyer-Kaiser and Calvin Mires coined the term 'Maritime Heritage Ecology', which is a field that must be explored further in the context of DSM (Meyer-Kaiser & Mires, 2022; Mires et al., 2024). Understanding UCH, including intangible UCH, is also key to respecting international commitments to human rights (including intergenerational rights, the right of public participation, the right to culture, and Indigenous peoples' rights).

Bottom trawling provides an example of what could happen if effective regulations are not passed for DSM (Jarvis, 2024). There are documented cases where bottom trawling gear tore through shipwrecks, destroying UCH and fish habitats, and this happens in areas where shipwrecks are not protected, or where bottom trawling is not regulated. While there are many unknowns regarding ocean ecosystems, especially in the deep, what is known shows the far-reaching impacts that humans can have. We know enough from damaging practices, like bottom trawling, to tell us that similar practices, like DSM, should not go ahead in the absence of robust and functional protections.

DSM is a real threat to Ocean Heritage. With metals prices plummeting and the very real threat of legal and financial liability costs, DSM may well not result in financial gains. Scientists are raising the alarm that DSM will have devastatingly wide impacts on microorganisms, fisheries, food chains, cultural traditions and more by impacting the seafloor and water column. However, for geopolitical reasons, and the opportunity for a small few to make money whether or not the industry (or humankind) as a whole does, there is a very real possibility that the industry will go ahead and without adequate protections in place for Ocean Heritage. In fact, test mining is already occurring.

Archaeologists and other ocean professionals cannot dismiss DSM as a potential threat, and their expertise is needed. Internationally, the ISA currently meets at least twice a year in Kingston, Jamaica, to negotiate regulations to govern DSM. UCH has only been given significant discussion in the past couple of years, and there is a dearth of maritime archaeologists on hand. Practical expertise is needed to elaborate on the type of information needed in surveys to identify areas of concern for Ocean Heritage, as well as the requisite monitoring and other technology that could potentially identify and protect such heritage from being destroyed. The regulations remain incomplete, and major unresolved issues linger. A 2024 peer reviewed paper found that 30 major issues in the ISA regulations remain outstanding (Pickens et al., 2024) and that the ISA internal 2025 target date for completion is unrealistic. It is not too late to get involved. Archaeologists can provide necessary information to the ISA via engagement with a member country or an Observer organisation, or by applying for their own organisation to be an Observer at the ISA.

At the national level, DSM is being proposed in national waters by Norway, Papua New Guinea, India, Japan, the Cook Islands, and the United States. In Papua New Guinea, Indigenous people's cultural connection to the ocean is at the forefront of opposition to DSM (Childs, 2019), but it is not apparent that maritime archaeologists are involved in the evaluation of many of those proposals.

This book serves as a call to action to familiarise oneself with the risks and the situation at hand to ensure that if DSM goes forward on any scale (including test mining) that Ocean Heritage is protected. The time is right, with the BBNJ Agreement on the horizon, a new Secretary General at the helm of the ISA, increasing interest from countries to explore mining their own seabed, and the UN Decade of Ocean Science for Sustainable Development (2021–2030) as a backdrop, it is not too late to tune-in. Now is the time for engagement.

References

Childs, J. (2019). Deep sea mining threatens indigenous culture in Papua New Guinea. *The Conversation*. http://theconversation.com/deep-sea-mining-threatens-indigenous-culture-in-papua-new-guinea-112012.

Jarvis, C. (Ed.). (2024). *Threats to our ocean heritage: Bottom trawling*. Springer Nature Switzerland. https://doi.org/10.1007/978-3-031-57953-0

Mires, C., Haskell, B., & Meyer-Kaiser, K. (2024). Multi-use and multi-vocal challenges of preserving underwater cultural heritage in Stellwagen Bank National Marine Sanctuary. In C. Jarvis (Ed.), *Threats to our ocean heritage: Bottom trawling* (pp. 67–82). Springer Nature Switzerland. https://doi.org/10.1007/978-3-031-57953-0

Meyer-Kaiser, K. S., & Mires, C. H. (2022). Underwater cultural heritage is integral to marine ecosystems. *Trends in Ecology & Evolution, 37*(10), 815–818. https://doi.org/10.1016/j.tree.2022.06.014

Pickens, C., Lily, H., Harrould-Kolieb, E., Blanchard, C., & Chakraborty, A. (2024). From what-if to what-now: Status of the deep-sea mining regulations and underlying drivers for outstanding issues. *Marine Policy, 169*, 105967. https://doi.org/10.1016/j.marpol.2023.105967

Appendix A: International Seabed Authority Issued Mining Contracts as of Publication

Exploration Contracts

The International Seabed Authority has entered into 15-year contracts for the exploration of minerals in the deep seabed with 22 contractors according to their website: https://www.isa.org.jm/exploration-contracts/.

Contractor	Contracts			
	Polymetallic nodules	Polymetallic sulphides	Cobalt-rich ferromanganese crusts	Total
Interoceanmetal Joint Organization	1			1
JSC Yuzhmorgeologiya	1			1
Government of the Republic of Korea	1	1	1	3
China Ocean Mineral Resources Research and Development Association	1	1	1	3
Deep Ocean Resources Cevelopment Co. Ltd.	1			1
Institut français de recherche pour l'exploitation de la mer	1	1		2
Nauru Ocean Resources Inc.	1			1
Tonga Offshore Mining Limited	1			1
Global Sea Mineral Resources NV	1			1
Loke CCZ (formerly UK Seabed Resources Ltd.)	2			2

(continued)

C. Jarvis (ed.), *Threats to Our Ocean Heritage: Deep Sea Mining*,
SpringerBriefs in Archaeology, https://doi.org/10.1007/978-3-031-98238-5

(continued)

Contractor	Contracts			
	Polymetallic nodules	Polymetallic sulphides	Cobalt-rich ferromanganese crusts	Total
Marawa Research and Exploration Ltd.	1			1
Ocean Mineral Singapore Pte. Ltd.	1			1
Cook Islands Investment Corporation	1			1
China Minmetals Corporation	1			1
Beijing Pioneer Hi-Tech Development Corporation	1			1
Ministry of Natural Resources and Environment of the Russian Federation		1	1	2
Government of Poland		1		1
Japan Organization for Metals and Energy Security			1	1
Companhia de Pesquisa de Recursos Minerais S.A. (renounced rights in relation to contract 28 December 2021)			1	1
Blue Minerals Jamaica Ltd.	1			1

Seventeen of these contracts are for the exploration of the Clarion-Clipperton Zone and will focus on polymetallic nodules. Other nodule contracts include one in the Central Indian Ocean Basin and one in the Western Pacific Ocean.

There are seven contracts for the exploration of polymetallic sulphides in the Southwest Indian Ridge, the Central Indian Ridge, and the Mid-Atlantic Ridge.

There are five contracts for the exploration of cobalt-rich crusts in the Western Pacific Ocean.

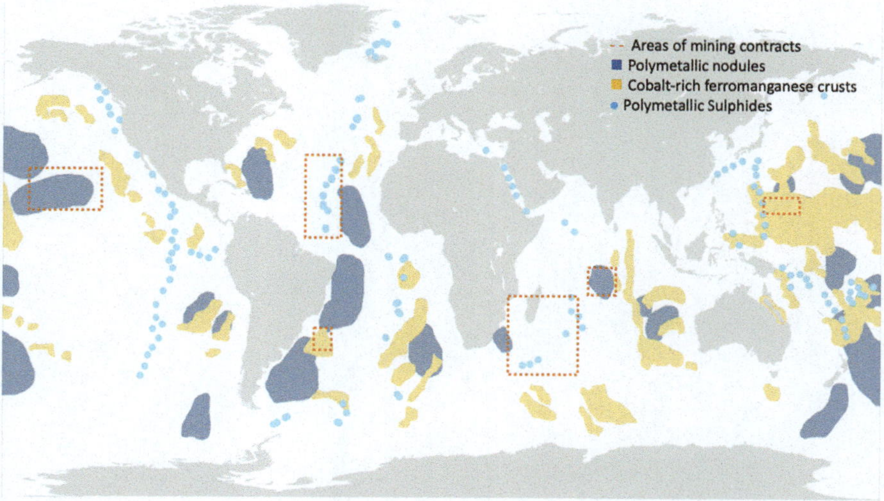

Source: Created by Charlotte Jarvis, 2024. Based on information from the International Seabed Authority

As per the ISA requirements:

Prior to the commencement of its programme of activities under the contract, each contractor is also required to submit to the Secretary-General a contingency plan to respond effectively to incidents arising from its activities in the exploration area. Each contractor is required to propose a programme for the training of nationals of developing States. As agreed with ISA, the training programme is incorporated into the contract as Schedule 3. Each contractor is also required to submit an annual report on its programme of activities.

Appendix B: Intangible Cultural Heritage in the Pacific and at the International Seabed Authority, Edwin 'Ekolu' Lindsey III

Words transcribed from Edwin 'Ekolu' Lindsey III, co-founder and president of Maui Cultural Lands.

'On behalf of the island voices from Hawai'i and our family members from Oceania, we first express our sincere appreciation to the indigenous people, residents and the government of Jamaica for allowing us to be hosted on your beautiful island home. Hawai'i shares our true spirit of Aloha for your losses and challenges that resulted from natures rage.

To the states gathered here we extend our greetings of Aloha to all of you from the people of Hawai'i. We especially give our thanks to Clement Yow Mulalap for his excellent facilitation in the intersession working group on cultural heritage… We want to particularly highlight our proposal for a committee on intangible underwater cultural heritage within the ISA.

We also wish to highlight that we also made a submission to the special rapporteur on climate change and human rights on deep sea mining and indigenous peoples. Our recommendations included enhancing transparency, strengthening mechanisms for free and prior informed consent, promotion of indigenous knowledge, protection of intangible underwater cultural heritage and international cooperation.

We welcome the United Kingdom observation that a cessation of activities would be appropriate while discussions are ongoing.

At the Festival of Pacific Arts and Culture, known as FESTPAC, In June 2024, the Tuurama Ariki Declaration was signed. The Tuurama Ariki Declaration signing saw leaders promising to utilise indigenous traditional knowledge to lead the way forward in communities of Oceania.

That Declaration stated that traditional Leaders will amplify our united and collective voices, exchange experiences and collaborate on solutions, advocate on issues that impact our region and our planet and respond to urgent challenges that affect Pacifika nations.

C. Jarvis (ed.), *Threats to Our Ocean Heritage: Deep Sea Mining*,
SpringerBriefs in Archaeology, https://doi.org/10.1007/978-3-031-98238-5

The word "Tuurama" *or the Hawaiian term* hōʻike i ka mālamalama *sheds light on ancestral wisdom guiding the journey ahead.*

Mr President [of the ISA], *it is our duty to protect our cultural heritage, our traditional knowledge, and our collective ancestral wisdom connected to the deep sea. We think that an important way of facilitating this ongoing duty is to establish the Committee on intangible underwater cultural heritage within the ISA.*

Another important way is to ensure that the regulations protect both our cultural heritage and our right of free, prior, and informed consent. We do want to emphasise that intangible heritage must not be subservient to tangible heritage. The protection of intangible heritage is foundational. Creation begins at the bottom of the sea, and is related to wali wali *or genetics, and from this intangible or genetic soup, all things are created. It is that part that we talk about when we talk about intangible cultural heritage which led to the creation of tangible things, and ultimately the creation of life itself. When we discuss tangible and intangible, we are in that sense talking about the same thing.*

Lastly Mr. President, we are very pleased to share that our voices here at the ISA council are not without support. Following the recent gathering of the Pacific Festival of Arts and Culture, the Governor of Hawaiʻi, on July 8, 2024, signed into law the Hawaiʻi Seabed Mining Prevention Act 228 which protects the waters of Hawaiʻi from the detrimental impacts of deep sea mining and related industries and more importantly, acknowledges the significance and importance of the indigenous cultural connections to the deep sea'.

Remarks given by: Edwin 'Ekolu' Lindsey III, co-founder and president of Maui Cultural Lands on July 15, 2024 at the International Seabed Authority Council.

Appendix C: Intangible Cultural Heritage in the Pacific and at the International Seabed Authority, Solomon 'Uncle Sol' Kaho'ohalahala

Words transcribed from Solomon 'Uncle Sol' Kaho'ohalahala, a seventh-generation resident of the island of Lāna'I and the Native Hawaiian Elder representative on the advisory council of the Papahānaumokuākea Marine National Monument.

Fig. C.1 Solomon 'Uncle Sol' Kaho'ohalahala speaks at an event held by the Te Ipukarea Society on the importance of intangible heritage and Indigenous experiences with the ocean. Photo credit: Bobbi-Jo Dobush, 2024

C. Jarvis (ed.), *Threats to Our Ocean Heritage: Deep Sea Mining*, SpringerBriefs in Archaeology, https://doi.org/10.1007/978-3-031-98238-5

'It is with humility and a sincere heartfelt "Spirit of Aloha", that I express our appreciation and privilege to address you, Mr. President, and the 168 countries represented in this esteemed Assembly of the ISA. I am very humbled and honoured to be here from Hawai'i, along with other observers from across the Pacific to offer our views about the importance and our cultural connections to the deep ocean.

When Queen Lili'uokalani, our last reigning queen, was overthrown in an illegal act of war, by US military forces in 1893 at 'Iolani Palace, the Hawaiian Kingdom as a sovereign and independent country was denied participation in the international assembly of nation states. As the course of history will show, Hawai'i has never been seated or represented at the table where the rules and regulations have been created by nation states, like the Law of the High Seas.

I would note that in the darkest hours of her life where her nation and her kingdom had just been stolen from her, while confined and imprisoned in her room, our Queen chose to translate "The Kumulipo", our genealogy chant of creation, into the English language. Our queen chose to do this—saying perhaps there will be a need for this in the future—and scientists and genealogists might want to take a look at this. It's now been 130 years since the translation of the Kumulipo and the overthrow. Perhaps her vision has relevancy for these critical climate change times and may be what we need to take pause to look back to our past so that we may move forward.

We have a Hawaiian proverb that says, I ka wa mamua, i ka wa mahope*—to advance, to innovate, to discover new solutions, and move forward, we must take pause, to look backward, to look to our past and to learn from those past experiences. Therefore, in looking backward, The Kumulipo provides for our connection to the deep seas and the island Earth.*

The Kumulipo is our Hawaiian creation chant, consisting of more than 2,100 lines that explain the beginning genealogy of all creatures. The Kumulipo begins with the cosmic darkness. The first lines of the chant tells us that we come from the deep seas, created from the "walewale"*, synonymous with the primordial sludges, the first creature to emerge is the* "Uku ko'a ko'a"*, or the first coral polyp. It continues with the creatures that come up through the vertical water column into nearshore waters, the low-lands, to the hills and the mountains, and even taking flight into the sky—are all part of our creation genealogy.*

We as a people, don't come into being until much much later in the creation of this Honua*, or island Earth. When we are ushered into this place we call home, the earth is in balance with the connection of sea, land, humankind, the gods and the* ponahakeola*, the circle of life. In our world view, we therefore bear the inherent responsibility to care for all creatures that precede us, of which, the coral polyp of the deep sea is our eldest ancestor.*

The proposed process to mine and dredge the seafloor is to enter the realm of our creation and an intrusion into the most sacred place of creation. It is perhaps the ultimate kind of destruction with no known outcomes and consequences for the great ocean that is our sacred place of creation, our country and our home.

I challenge the perception that culture can only be tangible artefacts. As seafaring people, we have travelled across the vast Pacific, the largest ocean on earth. Our observations are intimate experiences and knowledge collected and committed to memory over generations. We know the consistent patterns of the constellations, the phases of the moon, and seasons of the sun. We know the currents, the character and directional movements of waves, the clouds that foretell velocity and direction of the winds and impending weather systems. We know the resources that provide sustenance derived from the deep seas. We are the culture of the people of the deep sea and our culture is intangible. The ocean is everything to us. It's what we connected and depend upon for our sustenance and our survivability.

It's unacceptable to proceed with activities that will disrupt the most sacred place of our creation in the deep sea, ultimately disrupting the food chain which now we have subsisted on for millenia. In the conversations of deep sea mining, there is no consideration that we come from this space. It's about minerals. It's about resources. It's about extraction. It's about investments and profits with no regard for culture. It's that that bothers me about this process of deep sea mining.

I am the culture of the deep seas. I am the voice of our ancestors. I am here to say we come from this place. It is my home. As an indigenous Hawaiian I strongly urge this esteemed Assembly to consider our indigenous perspective. I oppose any form of deep sea mining that will destroy our culture, our home, our ocean, and our island Earth. Therefore, I call for indigenous voices to be included, acknowledged and incorporated in the decisions being considered in this Assembly.

I would like to end by calling our ancestral spirits to also be assembled into this important space and proceedings.

Solomon 'Uncle Sol;' Pili Kaho'ohalahala on July 26, 2023, at the International Seabed Authority Assembly day 3.

The manufacturer's authorised representative in the EU is Springer Nature Customer Service Centre GmbH, Europaplatz 3, 69115 Heidelberg, Germany. If you have any concerns regarding our products, please contact ProductSafety@springernature.com

Printed and bound by CPI Group (UK) Ltd, Croydon, CR0 4YY
05/10/2026
02228152-0011